高等院校信息技术规划教材

Oracle 数据库实验教程

李然 林远山 编著

清华大学出版社
北京

内容简介

本书是作者在长期从事数据库课程教学和科研的基础上,为满足教学需求而编写的配套实验指导书。本书以 Oracle 技术为基础,编排典型实验,能充分满足课程教学需求。实验内容通用;实验安排可操作性强;灵活性高。通过实验并结合典型系统进行分析,使学生较为系统地掌握 Oracle 数据库的基本开发方法,设计出满足一定规范的 Oracle 数据库应用系统。

本书共 7 章,18 个实验,基本内容包括 Oracle 10g 的基本安装、配置、卸载;Oracle 的体系结构;PL/SQL 程序设计;安全管理;数据库备份与恢复;手工建立数据库;PowerDesigner 的使用;项目综合实践。

实验内容循序渐进、深入浅出,可作为计算机科学与技术、软件工程、网络工程等相关专业大学本科和研究生相关课程的实验教材,同时也可以供参加自学考试的人员及数据库应用系统开发设计人员等阅读参考。

本书封面贴有清华大学出版社防伪标签,无标签者不得销售。
版权所有,侵权必究。举报: 010-62782989, beiqinquan@tup.tsinghua.edu.cn。

图书在版编目(CIP)数据

Oracle 数据库实验教程/李然,林远山编著. —北京: 清华大学出版社,2016(2024.9重印)
(高等院校信息技术规划教材)
ISBN 978-7-302-45031-3

Ⅰ. ①O… Ⅱ. ①李… ②林… Ⅲ. ①关系数据库系统-高等学校-教材 Ⅳ. ①TP311.138

中国版本图书馆 CIP 数据核字(2016)第 218476 号

责任编辑: 张 玥 赵晓宇
封面设计: 常雪影
责任校对: 梁 毅
责任印制: 曹婉颖

出版发行: 清华大学出版社
网　　址: https://www.tup.com.cn, https://www.wqxuetang.com
地　　址: 北京清华大学学研大厦 A 座　　　邮　编: 100084
社 总 机: 010-83470000　　　　　　　　　　邮　购: 010-62786544
投稿与读者服务: 010-62776969, c-service@tup.tsinghua.edu.cn
质量反馈: 010-62772015, zhiliang@tup.tsinghua.edu.cn
课件下载: https://www.tup.com.cn, 010-83470236
印 装 者: 河北盛世彩捷印刷有限公司
经　　销: 全国新华书店
开　　本: 185mm×260mm　　　印　张: 10.5　　　字　数: 249 千字
版　　次: 2016 年 10 月第 1 版　　　　　　　　印　次: 2024 年 9 月第 4 次印刷
定　　价: 39.50 元

产品编号: 067852-02

前言

Oracle 数据库是一款关系数据库管理系统,在数据库领域一直处于领先地位。Oracle 数据库系统是目前世界上流行的关系数据库管理系统,系统可移植性好、使用方便、功能强,适用于各类大、中、小、微机环境,是一种高效率、可靠性好的适应高吞吐量的数据库解决方案。

本书针对大型数据库 Oracle 实践要求高,实验指导教材少的特点,主要以数据库管理系统 Oracle 10g 和数据库设计工具 PowerDesigner 为平台,通过大量的实验培养学生对数据库的基本操作、设计、开发及维护的能力。

本书共 7 章,18 个实验,基本内容如下:

第 1 章数据库起步,主要内容包括 Oracle 10g 的基本安装方法;监听配置、网络服务配置;Oracle 10g 的卸载方法;Oracle 数据库启动与关闭。

第 2 章 Oracle 的体系结构,主要内容包括 Oracle 数据库的物理存储结构;Oracle 数据库的逻辑存储结构。

第 3 章 PL/SQL 程序设计,主要内容包括 SQL*Plus 环境命令的使用;PL/SQL 基本编程方法。

第 4 章安全管理,主要内容包括 Oracle 数据库用户的创建与维护;Oracle 数据库权限的分配与管理;Oracle 数据库角色的创建与管理;Oracle 资源配置文件 profile 的创建与管理;触发器的使用方法;审计设置、查看的方法。

第 5 章数据库备份与恢复,主要内容包括冷备份与恢复;日志文件的管理;归档模式的管理;归档模式下的数据库备份与恢复;逻辑备份;数据闪回。

第 6 章手工建立数据库,主要内容包括手工建立数据库的方法;PowerDesigner 的使用。

第 7 章项目实践,主要内容包括项目实践按照软件工程开发过程,通过需求分析、概要设计、逻辑设计、物理设计、数据库实施等阶段完成设计。

本书由李然编写,林远山整理。在编写过程中参阅了大量的参考书目和文献资料。本书的出版也得到了清华大学出版社的大力支持,责任编辑张玥为本书付出了辛勤的劳动,在此一并表示衷心的感谢。

由于作者水平有限,书中难免有错误,敬请读者批评指正。作者的邮箱是 liran@dlou.edu.cn。

编　者
2016 年 9 月

目录

第 1 章 数据库起步 ... 1
实验 1.1　Oracle 安装及卸载 ... 1
实验 1.2　Oracle 数据库启动与关闭 ... 11

第 2 章 Oracle 的体系结构 ... 15
实验 2　Oracle 的体系结构 ... 15

第 3 章 PL/SQL 程序设计 ... 24
实验 3.1　SQL*Plus 命令的使用 ... 24
实验 3.2　PL/SQL 基本编程方法 ... 27

第 4 章 安全管理 ... 30
实验 4.1　用户、权限管理 ... 30
实验 4.2　触发器 ... 36
实验 4.3　审计 ... 41

第 5 章 数据库备份与恢复 ... 48
实验 5.1　冷备份与恢复 ... 48
实验 5.2　日志文件的管理 ... 54
实验 5.3　归档模式的管理 ... 57
实验 5.4　联机备份 ... 66
实验 5.5　归档模式下的数据库恢复 ... 72
实验 5.6　逻辑备份 ... 83
实验 5.7　数据泵 ... 87
实验 5.8　数据闪回 ... 94

第6章 手工建立数据库 ………………………………………………………………… 106

实验 6.1 手工建立数据库 ……………………………………………………… 106

实验 6.2 PowerDesigner 使用简介 …………………………………………… 121

第7章 项目实践 ………………………………………………………………………… 133

项目 7.1 考试报名管理系统 …………………………………………………… 133

项目 7.2 图书管理系统 ………………………………………………………… 138

附录 SQL＊Plus 环境命令 …………………………………………………………… 147

参考文献 …………………………………………………………………………………… 160

第1章 数据库起步

本章要求重点掌握如何配置数据库的连接,学习难点是数据库实例的启动与关闭。

实验 1.1 Oracle 安装及卸载

【实验目的】

(1) 掌握 Oracle 10g 的基本安装方法。
(2) 掌握 Oracle 10g 的监听配置、网络服务配置。
(3) 掌握 Oracle 10g 的卸载方法。

【实验内容】

(1) Oracle 10g 的安装。
(2) Oracle 10g 的监听、网络配置。
(3) Oracle 10g 的卸载。

【实验步骤】

目前,Oracle 10g 产品可以直接从 Oracle 的官方网站下载软件,网址是 http://www.Oracle.com/technology/software。官方免费软件与购买的正版软件是有区别的,主要区别在于 Oracle 能够支持的用户数量、处理器数量及磁盘空间和内存的大小。Oracle 提供的免费软件主要针对的是学生和中小型企业等,目的是使他们熟悉 Oracle,占领未来潜在的市场。另外,当安装完成 Oracle 以后,还需要到 OracleMetaLink 网站下载最新的补丁包,网址是 http://metalink.oracle.com。只有购买正版 Oracle 产品并获得授权许可的用户才拥有注册码并可以登录该网站。

若用户使用下载的软件,则不能得到 Oracle 公司的技术支持,而且一旦被发现使用未经授权许可的 Oracle 产品,由此所付出的代价比购买正版软件所付出的代价要大得多。同时,从 Oracle 官方网站的下载许可协议中也可以看到,以下载方式得到的软件产品只能用于学习和培训等,不得用于商业目的。所以,企业应使用正版的 Oracle 软件。

在安装 Oracle Database 之前,必须明确系统安装所需要的条件。

1. 硬件环境

安装 Oracle 10g 数据库的硬件需求如表 1-1 所示。

表 1-1　安装 Oracle 10g 数据库的硬件需求

需　求	最　小　值	需　求	最　小　值
物理内存	最小 256MB,推荐 512MB	硬盘空间	1.5GB
虚拟内存	物理内存的 2 倍	显卡	256 色
临时磁盘空间	100MB	处理器	最小 200MHz,推荐 550MHz 以上

在硬件需求中,处理器的速度和内存大小直接影响着 Oracle 运行的速度,所以建议硬件配置越高越好。一般地,服务器配置应高于客户端配置,而且配置越高安装速度越快。

2. 软件环境

安装 Oracle 10g 数据库的软件需求如表 1-2 所示。

表 1-2　安装 Oracle 10g 数据库的软件需求

需　求	最　小　值
系统体系结构	32 位
操作系统	Oracle Database 10g 支持广泛的操作系统平台,从 UNIX、Linux 到 Windows 等都针对不同操作系统提供了不同的版本。就 Oracle 10g for Windows 来说,支持 Windows 2000、Windows XP 及 Windows 2003,不再支持 Windows 98 和 Windows NT。由于不同操作系统在网络系统中扮演的角色不同,可以安装 Oracle 组件的内容也不同,因此 Oracle Database 10g 呈现的功能也有所不同。如果要建立 Oracle 数据库服务器,则必须将该软件安装在 Windows Server 2000/2003 文件服务器上
网络协议	Oracle Net 基础层使用 Oracle 协议支持与下列行业标准网络协议之间的通信: • TCP/IP; • 带有 SSL 的 TCPIP; • 命名管道
Web 浏览器	在 Oracle 10g 中,可用于 iSQL * Plus 和 Oracle Enterprise Manager Database Control 的浏览器有以下版本: • Netscape Navigator 7.2 以上的版本; • Microsoft Internet Explorer 6.0 以上的版本; • Mozilla 1.7 以上的版本; • Safari 1.2; • Firefox 1.0.4

仔细阅读安装要求,确保当前安装环境能够满足安装需求。如果系统曾安装过 Oracle,则卸载 Oracle_Home 环境变量。若当前已经安装了一个 Oracle 并想重新安装,则要卸载已安装的 Oracle。

在启动 Oracle Database 10g 安装程序时，系统自动检测环境是否满足安装要求，如果满足条件，则创建安装文件并继续安装，否则拒绝安装。

本实验以 Windows XP 为例，其他操作系统的安装与此类似。

下载 Oracle 10g 后，解压到一个文件夹下，双击 setup.exe 文件即可启动安装界面，如图 1-1 所示。

图 1-1　Oracle Universal Installer 启动安装界面

Oracle 主目录位置就是 Oracle 准备安装的位置，称为 Oracle_Home，一般 Oracle 根据当前计算机的硬盘大小默认给出一个合适的位置。安装 Oracle 时可以只安装 Oracle 软件，然后单独创建数据库，也可以在图 1-2 中选中"创建启动数据库（附加 720MB）"复选框，在安装 Oracle 产品的同时创建一个数据库。对于初学者来说，推荐这样安装。填写全局数据库名，以及管理员的密码。本例密码是 jsj，全局数据库名是数据库在服务器网络中的唯一标识。

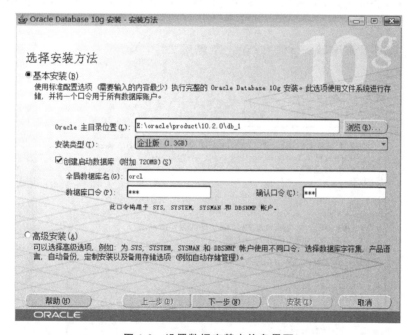

图 1-2　设置数据库基本信息界面

单击"下一步"按钮，就会出现产品特定的先决条件检查界面，如图 1-3 所示。开始对 Oracle 服务器进行环境检查，主要查看服务器是否符合 Oracle 安装的条件，如操作系统是否支持、系统内存是否符合 Oracle 安装的最低要求等。

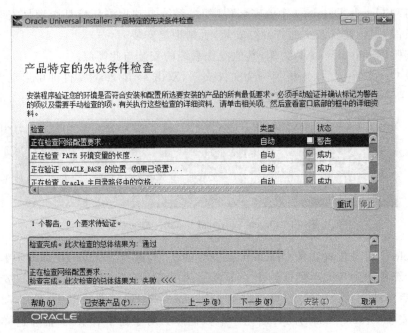

图 1-3 Oracle 安装前环境检查界面

Oracle 检查通过后,单击"下一步"按钮,就会列出所有安装 Oracle 过程中的默认选项,如图 1-4 所示。

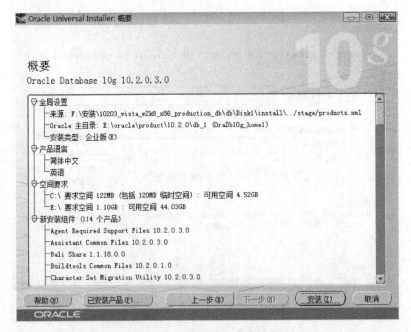

图 1-4 Oracle 默认安装设置

单击"安装"按钮,进入安装界面,这一过程经历时间比较长,根据计算机的性能不同有很大差别,如图 1-5 所示。

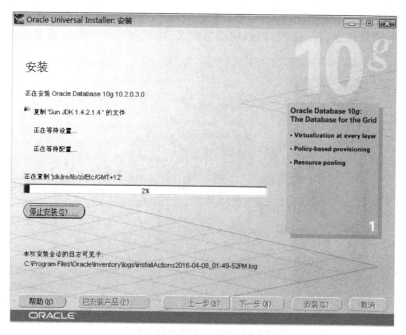

图 1-5　Oracle 安装界面

图 1-5 完成后,进入各种 Oracle 工具的安装阶段,包括网络配置向导、iSQL＊Plus 等,如图 1-6 所示。

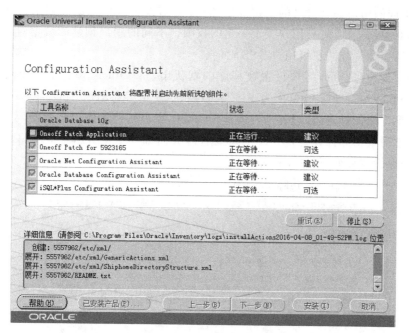

图 1-6　Oracle 各种工具的安装

接下来自动启动 DBCA(Database Configuration Assistant)进入创建默认数据库阶

段,如图 1-7 所示。Oracle 中的数据库主要包括存放数据的文件,这些文件在 Oracle 安装完成后,在计算机硬盘上都能找到,包括数据文件、控制文件和数据库日志文件。

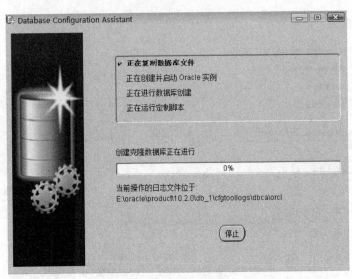

图 1-7　DBCA 创建数据库界面

虽然一个 Oracle 数据库服务器中可以安装多个数据库,但是由于一个数据库需要占用非常大的内存空间,因此一般一个服务器只安装一个数据库。每一个数据库可以有很多用户,不同的用户拥有自己的数据库对象(如数据库表),一个用户如果访问其他用户的数据库对象,必须由对方用户授予一定的权限。不同的用户创建的数据对象只能被数据对象的属主和系统的超级用户 SYS 访问。

数据库创建完毕后,需要设置数据库的默认用户。Oracle 预置了两个用户分别是 SYS 和 SYSTEM。同时 Oracle 为程序测试提供了一个普通用户 SCOTT,口令管理中可以对数据库用户设置密码,设置是否锁定。Oracle 客户端使用用户名和密码登录 Oracle 系统后才能对数据库操作。DBCA 的口令管理界面如图 1-8 所示。单击"口令管理"按钮,进入口令设置界面,如图 1-9 所示。

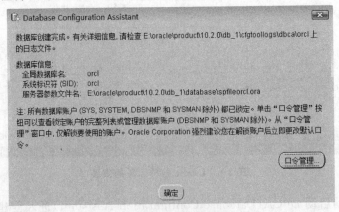

图 1-8　DBCA 口令管理界面

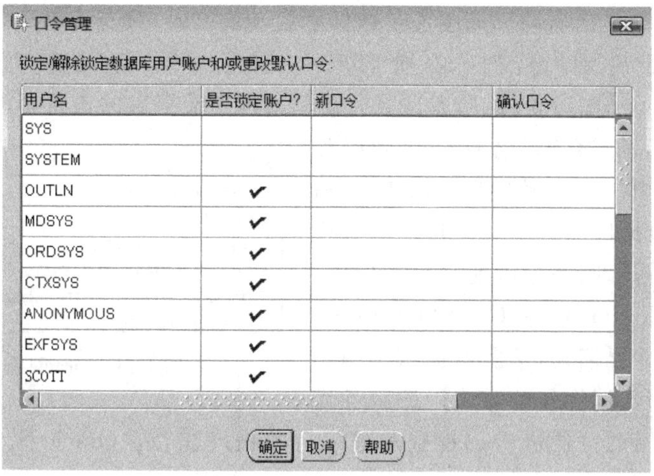

图 1-9 口令管理界面

默认的用户中，SYS 和 SYSTEM 用户是没有锁定的，安装成功后可以直接使用。SCOTT 用户默认为锁定状态，因此不能直接使用，需要把 SCOTT 用户设定为非锁定状态才能正常使用。

安装结束后会出现安装结束界面，将界面上的安装信息记录到文件中，对以后维护数据库非常有用。单击"退出"按钮结束安装，如图 1-10 所示。

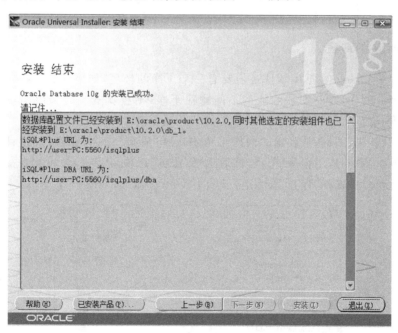

图 1-10 安装结束界面

安装完毕后，右击"计算机"图标，在弹出的快捷菜单中选择"管理"命令，打开"计算机管理"窗口，单击"服务和应用程序"节点，然后选择"服务"选项，可以看到，Oracle 服务

OracleServiceORCL 已启动。该服务是数据库启动的基础,只有该服务启动了,Oracle 数据库才能正常启动。这时,必须启动的服务和监听程序 OracleOraDb10g_home1TNSListener 已启动,该服务是服务器端为客户端提供的监听服务,只有该服务在服务器上正常启动,客户端才能连接到服务器。该监听服务接受客户端发出的请求,然后将请求传递给数据库服务器。一旦建立了连接,客户端和数据库服务器就能直接通信了。SQL*Plus 应用服务程序 OracleOraDb10g_home1iSQL*Plus 已启动,该服务提供了用浏览器对数据库中数据操作的方式。该服务启动后,就可以使用浏览器进行远程登录并进行数据库操作。Oracle 服务界面如图 1-11 所示。

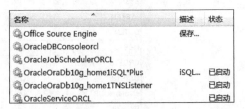

图 1-11 Oracle 服务界面

接下来验证数据库。选择"开始"→"所有程序"→"附件"→"命令提示符"命令,启动 DOS 命令窗口,连接数据库。使用超级用户 SYS,口令 jsj。

```
C:\Users\user>sqlplus/nolog
SQL*Plus: Release 10.2.0.3.0-Production on 星期五 4月 8 15:14:09 2016
Copyright (c) 1982, 2006, Oracle. All Rights Reserved.
SQL>conn sys/jsj as sysdba
已连接。
```

3. Oracle 的卸载

因为安装 Oracle 数据库软件必须要有一个干净的环境,如果安装失败,一定存在某些环境原因。例如,以前安装的软件不能删除干净,则重新安装时会出错,这时要卸载 Oracle,具体方法为:

右击"计算机"图标,从弹出的快捷菜单中选择"管理"命令,在打开的"计算机管理"窗口中选择"服务和应用程序"节点,然后选择"服务"选项,找到 Oracle 相关服务右击,从弹出的快捷菜单中选择"停止"命令,停止所有 Oracle 服务,如图 1-12 所示。

图 1-12 Oracle 服务界面

在程序组中的 Oracle Installation Products 中启动 Universal Installer,进入启动界

面,如图 1-13 和图 1-14 所示。

图 1-13　Universal Installer 启动界面

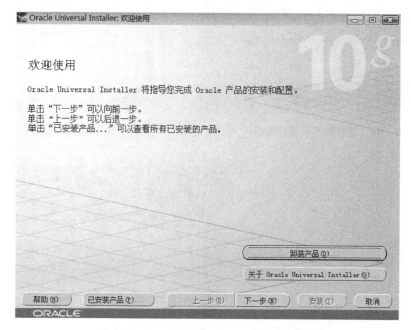

图 1-14　Oracle Universal Installer 欢迎界面

单击"下一步"按钮,进入卸载产品界面,如图 1-15 所示。选择要卸载的产品,单击"删除"按钮,系统进入卸载产品及相关组件界面,如图 1-16 所示。

单击"是"按钮,系统进入删除界面,如图 1-17 所示。系统将运行一段时间,将选中的产品删除。删除完成界面如图 1-18 所示。

由于 Oracle 本身的卸载软件不能完全卸载,因此要用手动删除的方式删除注册表信息。具体做法如下:

(1) 在操作系统界面上选择"开始"→"运行"命令,输入 regedit 后单击 Enter 键。

(2) 选择 HKEY_LOCAL_MACHINE\SOFTWARE\ORACLE,按 Delete 键删除这个入口。

(3) 选择 HKEY_LOCAL_MACHINE\SYSTEM\CurrentControlSet\Services,滚动这个列表,删除所有 Oracle 入口。

(4) 选择 HKEY_LOCAL_MACHINE\SYSTEM\CurrentControlSet\Services\Eventlog\Application,删除所有 Oracle 入口。

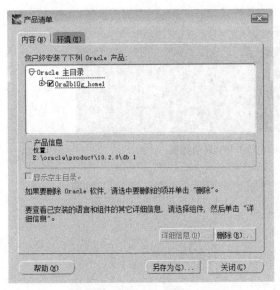

图 1-15 卸载的产品选择界面

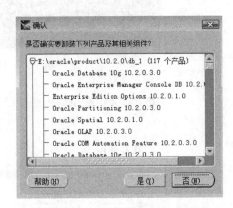

图 1-16 卸载产品及相关组件界面

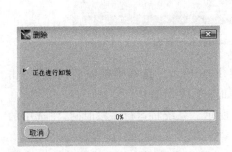

图 1-17 删除界面

图 1-18 删除完成界面

（5）选择"开始"→"控制面板"命令，在打开的窗口中双击"系统"选项，在"系统属性"对话框中选择"高级"选项卡，单击"环境变量"按钮，删除环境变量 CLASSPATH 和 PATH 中有关 Oracle 的设定。

（6）在操作系统界面上右击"我的电脑"图标，从弹出的快捷菜单中选择"属性"命令，出现"系统属性"对话框，单击"高级"选项卡中的"环境变量"按钮，删除环境变量 CLASSPATH 和 PATH 中有关 Oracle 的设定。

（7）从桌面上、STARTUP（启动）组、程序菜单中删除所有有关 Oracle 的组和图标。

（8）删除 Program Files\Oracle 目录。
（9）重新启动计算机，重启后才能完全删除 Oracle 所在的目录。

实验 1.2　Oracle 数据库启动与关闭

【实验目的】

（1）了解启动与关闭数据库的不同方式。
（2）掌握各种启动与关闭数据库的方法。
（3）了解各种启动与关闭数据库方法的适用条件。

【实验内容】

（1）各种启动数据库的方法。
（2）各种关闭数据库的方法。

【实验步骤】

1. 启动数据库

启动和关闭数据库必须具有 sysdba 权限，因此采用 SYS 用户登录。对于数据库的启动，有三种常用选项。

```
SQL> conn sys/jsj as sysdba
已连接。
```

（1）startup nomount。

首先从 spfile 或 pfile 中读取数据库参数文件，然后分配 SGA 和创建后台进程。这种方式下启动可执行：重建控制文件、重建数据库、修改初始化参数、查看部分动态性能视图（与内存和后台进程相关的视图）。

```
SQL> startup nomount
Oracle 例程已经启动。
Total System Global Area  612368384 bytes
Fixed Size                  1292036 bytes
Variable Size             192940284 bytes
Database Buffers          411041792 bytes
Redo Buffers                7094272 bytes
```

（2）startup mount。

在这个阶段，Oracle 根据参数文件（Pfile 或 Spfile）中的参数（CONTROL_FILES）找到控制文件（Control File），然后打开控制文件，加载控制文件到内存。从控制文件中获得数据文件和联机日志文件（Redo Log File）的名字及位置。这个时候，Oracle 已经把实例和数据库联系起来。对于普通用户来说,数据库还是不可访问。这个阶段主要用于数据库的维护,可执行:数据库日志归档、数据库介质恢复、数据文件联机或脱机、重新定位数据文件和重做日志文件、查看所有动态性能视图、修改数据库的归档模式、开启和关闭数据库闪回的功能。

startup mount 等于以下两个命令:

- startup nomount
- alter database mount

```
SQL> startup mount
Oracle 例程已经启动。
Total System Global Area  612368384 bytes
Fixed Size                  1292036 bytes
Variable Size             197134588 bytes
Database Buffers          406847488 bytes
Redo Buffers                7094272 bytes
数据库装载完毕。
```

(3) startup open。

数据库打开默认方式,可以不加 open。数据库可以进行正常的操作处理,主要是打开控制文件、数据库文件和日志文件。

startup open 等于以下三个命令:

- startup nomount
- alter database mount
- alter database open

```
SQL> startup open
Oracle 例程已经启动。
Total System Global Area  612368384 bytes
Fixed Size                  1292036 bytes
Variable Size             201328892 bytes
Database Buffers          402653184 bytes
Redo Buffers                7094272 bytes
数据库装载完毕。
数据库已经打开。
```

有些操作只能在 mount 下和 nomount 下完成 例如改归档模式就只能在 mount 下。open 并不是一切皆可以做。nomount、mount、open 没有子集包含的关系。

(4) startup force。

强制启动方式,当不能关闭数据库时可以用 startup force 来完成数据库的关闭,先关闭数据库,再执行正常启动数据库命令。

(5) startup pfile=参数文件名。

带初始化参数文件的启动方式,先读取参数文件,再按参数文件中的设置启动数据库。

2. 数据库的关闭

对于数据库的关闭,有 4 种不同的关闭选项。

(1) shutdown normal。

这是数据库关闭 shutdown 命令的默认选项。发出该命令后,任何新的连接都将不再允许连接数据库。在数据库关闭之前,Oracle 将等待现在连接的任何用户都从数据库中退出后才开始关闭数据库。采用这种方式关闭数据库,在下一次启动时无需进行任何的实例恢复。但需要注意的一点是,采用这种方式,也许关闭一个数据库需要几天时间,或更长。

```
SQL> shutdown normal
数据库已经关闭。
已经卸载数据库。
Oracle 例程已经关闭。
```

举例:相当于商店采用如下方式关门:①顾客出了门就不能再进来了;②不撵里边的顾客,等他们自愿地全走完,商店才关门。

(2) shutdown immediate。

这是一种常用的关闭数据库的方式,采用这种方式可以很快地关闭数据库。当前正在被 Oracle 处理的 SQL 语句立即中断,系统中任何没有提交的事务全部回滚。假如系统中存在一个很长的未提交的事务,采用这种方式关闭数据库也需要一段时间(该事务回滚时间)。系统不等待连接到数据库的任何用户退出系统,强行回滚当前任何的活动事务,然后断开所有的连接用户。

```
SQL> shutdown immediate
数据库已经关闭。
已经卸载数据库。
Oracle 例程已经关闭。
```

举例:相当于商店采用如下方式关门:①新顾客不能进入商店,在店内的顾客立刻终止选购商品,将商品放回货架,然后离开;②待顾客全部离开后关门。

(3) shutdown transactional。

该命令常用来计划关闭数据库,它使当前连接到系统且正在活动的事务执行完毕,运行该命令后,任何新的连接和事务都是不允许的。在所有活动的事务完成后,数据库将以 shutdown immediate 同样的方式关闭数据库。

```
SQL> shutdown transactional
数据库已经关闭。
已经卸载数据库。
Oracle 例程已经关闭。
```

举例:相当于商店采用如下方式关门:①出去的顾客不让再进入,新顾客不能进入商店；②在店内的顾客买完正在选购的商品后，不能再买其他商品，即离开商店；③待商店的顾客都离开后，商店关门。

（4）shutdown abort。

这是关闭数据库的最后一招，也是在没有任何办法关闭数据库的情况下才不得不采用的方式，一般不要采用。假如下列情况出现时，可以考虑采用这种方式关闭数据库。①数据库处于一种非正常工作状态，不能用 shutdown normal 或 shutdown immediate 这样的命令关闭数据库；②需要立即关闭数据库；③在启动数据库实例时碰到问题。

```
SQL> shutdown abort
Oracle 例程已经关闭。
```

举例：相当于商店采用如下方式关门：商店内的顾客将商品扔掉后立刻离开，可能有的顾客还没有离开，商店就已经关门了。

第 2 章 Oracle 的体系结构

从存储结构的角度来说，Oracle 数据库可以分为物理结构和逻辑结构。物理结构就是构成数据库的各种磁盘文件，是数据库的物理载体，这些物理文件包括存储数据库中的所有数据信息的数据文件，维护数据库的全局物理结构的控制文件，记录数据的改变、用于数据恢复的日志文件。逻辑结构主要有表空间、段、区间等。

实验 2 Oracle 的体系结构

【实验目的】

(1) 了解 Oracle 数据库的物理存储结构。
(2) 了解 Oracle 数据库的逻辑存储结构。
(3) 了解 Oracle 实例。

【实验内容】

(1) 查看 Oracle 的物理文件。
(2) 建立表空间。
(3) 维护表空间。

【实验步骤】

Oracle 数据库主要的物理存储结构包括构成数据库的各种物理文件，包括数据文件、控制文件、重做日志文件、归档重做日志文件、参数文件、警告、跟踪日志文件。

1. 查看 Oracle 的物理文件的信息

使用 SQL 命令从数据字典视图 v＄datafile 中查找数据文件信息。

```
SQL> col name for a55
SQL> select status,bytes,name from v$ datafile;
```

```
STATUS      BYTES NAME
------- ---------- ------------------------------------------------------
SYSTEM  524288000 E:\ORACLE\PRODUCT\10.2.0\ORADATA\ORCL\SYSTEM01.DBF
ONLINE   73400320 E:\ORACLE\PRODUCT\10.2.0\ORADATA\ORCL\UNDOTBS01.DBF
ONLINE  304087040 E:\ORACLE\PRODUCT\10.2.0\ORADATA\ORCL\SYSAUX01.DBF
ONLINE    5242880 E:\ORACLE\PRODUCT\10.2.0\ORADATA\ORCL\USERS01.DBF
ONLINE  104857600 E:\ORACLE\PRODUCT\10.2.0\ORADATA\ORCL\EXAMPLE01.DBF
ONLINE    2097152 E:\ORACLE\PRODUCT\10.2.0\ORADATA\ORCL\USER01.DBF
ONLINE     761856 E:\ORACLE\PRODUCT\10.2.0\ORADATA\ORCL\TEST.DBF
```

查询结果表明，orcl 数据库共有 7 个数据文件。

使用 SQL 命令从数据字典视图 v$controlfile 中查找控制文件信息。

```
SQL> select name,FILE_SIZE_BLKS from v$controlfile;

NAME                                                      FILE_SIZE_BLKS
--------------------------------------------------------- --------------
E:\ORACLE\PRODUCT\10.2.0\ORADATA\ORCL\CONTROL01.CTL                  464
E:\ORACLE\PRODUCT\10.2.0\ORADATA\ORCL\CONTROL02.CTL                  464
E:\ORACLE\PRODUCT\10.2.0\ORADATA\ORCL\CONTROL03.CTL                  464
```

查询结果表明，orcl 数据库共有三个镜像的控制文件。

使用 SQL 命令从数据字典视图 v$logfile 中查找联机日志文件信息。

```
SQL> select group#, status, type, member from v$logfile;
GROUP#  STATUS   TYPE    MEMBER
------- -------- ------- -----------------------------------------------
     3           ONLINE  E:\ORACLE\PRODUCT\10.2.0\ORADATA\ORCL\REDO03.LOG
     2  STALE    ONLINE  E:\ORACLE\PRODUCT\10.2.0\ORADATA\ORCL\REDO02.LOG
     1  STALE    ONLINE  E:\ORACLE\PRODUCT\10.2.0\ORADATA\ORCL\REDO01.LOG
```

查询结果表明，orcl 数据库共有三个重做日志组，其组号分别为 1、2、3，每组只有一个成员，且都处在联机状态。STATUS 字段为空白，表示文件正在使用；为 STALE 时表示该文件的内容是不完全的。

在操作系统环境下，同样可以看到 orcl 的物理结构，如图 2-1 所示。

查看参数文件 pfile。Oracle 的参数文件主要有两个：pfile 和 spfile。pfile 默认的名称为 init.ora，文件路径为 E:\Oracle\product\10.2.0\admin\orcl\pfile，这是一个文本文件，可以用任何文本编辑工具打开。

spfile 默认的名称为 spfile+例程名.ora，文件路径为 E:\Oracle\product\10.2.0\db_1\database，以二进制文本形式存在，不能用文本编辑器对其中参数进行修改。

两个文件可以用命令 CREATE PFILE FROM SPFILE 或 CREATE SPFILE FROM PFILE 来互相创建。

pfile 和 spfile 的区别：

（1）启动次序 spfile 优先于 pfile。

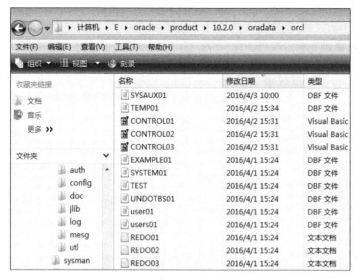

图 2-1　orcl 的物理结构

（2）pfile 是静态文件，修改之后不会马上生效，数据库必须重新启动读取这个文件才生效。

（3）spfile 是动态参数文件，是二进制文件，不可以直接用记事本等程序做修改，可以用 alter 命令做修改，不用重启数据库也能生效。

可使用 SQL 命令修改动态参数：

```
alter system set 参数名=值 scope=参数
```

参数取值有以下三种：

- scope=spfile。对参数的修改记录在服务器初始化参数文件中，修改后的参数在下次启动 db 时生效。适用于动态和静态初始化参数。
- scope=memory。对参数的修改记录在内存中，对于动态初始化参数的修改立即生效。在重启 db 后会丢失，会复原为修改前的参数值。
- scope=both。对参数的修改会同时记录在服务器参数文件和内存中，对于动态参数立即生效，对于静态参数不能用这个选项。

如果使用了服务器参数文件 spfile，则在执行 alter system 语句时，scope=both 是 default 的选项。

如果没有使用服务器参数文件，在执行 alter system 语句时指定 scope=spfile|both 会出错。

使用 show parameter 可以在 SQL * Plus 环境下查看数据库参数，也可以用记事本打开 init.ora 查看部分参数。如查看当前数据库名：

```
SQL> show parameter db_name
NAME                                 TYPE        VALUE
------------------------------------ ----------- -----
```

db_name   string   orcl

2. 表空间管理

表空间(Tablespace)对应一个或多个数据文件,表空间的大小是它所对应的数据文件大小的总和。Oracle 数据库中的数据逻辑存储在表空间中,物理存储在数据文件中。

使用 create tablespace 命令创建一个表空间 dltest。

```
SQL> create tablespace dltest
 2   datafile 'e:\Oracle\product\10.2.0\oradata\orcl\dltest_01.dbf' size 100k reuse
 3   autoextend on next 100k maxsize 2m,
 4   'e:\Oracle\product\10.2.0\oradata\orcl\dltest_02.dbf' size 100k reuse
 5   online
 6   permanent;
```

表空间已创建。

dltest 表空间包含两个数据文件 dltest_01.dbf 和 dltest_02.dbf。dltest_01.dbf 文件的初始大小是 100KB,如果该表空间已满或数据对象要求该表空间有更多的空间,它能自动扩展,每次扩展的大小是 100KB,最大可增到 2MB。dltest_02.dbf 文件的初始大小是 100KB,不能自动扩展。两个数据文件初始状态均为联机,并且是永久表空间,永久表空间存放的是永久对象,这是相对临时表空间而言。reuse 选项的作用是当需要添加新的数据文件,而数据文件已存在时,重用该文件。

使用 SQL 命令在数据字典 dba_data_files 中查看表空间及其包含的数据文件。

```
SQL> col tablespace_name for a10
SQL> col file_name for a55
SQL> col autoextensible for a4
SQL> set linesize 500
SQL> select tablespace_name,bytes,autoextensible,file_name from dba_data_
files;
```

TABLESPACE	BYTES	AUTO	FILE_NAME
USERS	5242880	YES	E:\ORACLE\PRODUCT\10.2.0\ORADATA\ORCL\USERS01.DBF
SYSAUX	304087040	YES	E:\ORACLE\PRODUCT\10.2.0\ORADATA\ORCL\SYSAUX01.DBF
UNDOTBS1	73400320	YES	E:\ORACLE\PRODUCT\10.2.0\ORADATA\ORCL\UNDOTBS01.DBF
SYSTEM	524288000	YES	E:\ORACLE\PRODUCT\10.2.0\ORADATA\ORCL\SYSTEM01.DBF
EXAMPLE	104857600	YES	E:\ORACLE\PRODUCT\10.2.0\ORADATA\ORCL\EXAMPLE01.DBF
USER01	2097152	NO	E:\ORACLE\PRODUCT\10.2.0\ORADATA\ORCL\USER01.DBF
TEST	761856	YES	E:\ORACLE\PRODUCT\10.2.0\ORADATA\ORCL\TEST.DBF
DLTEST	106496	YES	E:\ORACLE\PRODUCT\10.2.0\ORADATA\ORCL\DLTEST_01.DBF
DLTEST	106496	NO	E:\ORACLE\PRODUCT\10.2.0\ORADATA\ORCL\DLTEST_02.DBF

已选择 9 行。

查询结果表明,dltest 表空间对应的两个数据文件已建成。

创建 undo 表空间,undo 表空间用于保存回滚数据库变化所需的信息。

sql> create undo tablespace undotbs2 datafile 'e:\Oracle\product\10.2.0\oradata\
orcl\undo02.dbf' size 2m reuse autoextend on;
表空间已创建。

创建临时表空间 dltemp,临时表空间作为排序使用。

SQL> create temporary tablespace dltemp tempfile 'e:\Oracle\product\10.2.0\orada
ta\orcl\dltemp01.dbf' size 50m autoextend on next 50m maxsize 100m;
表空间已创建。

使用 SQL 命令在数据字典 dba_data_files 中查看表空间信息。

SQL> select tablespace_name,bytes,autoextensible,file_name from dba_data_files;

```
TABLESPACE   BYTES AUTO FILE_NAME
----------   -------- ---- ------------------------------------------------
USERS        5242880 YES  E:\ORACLE\PRODUCT\10.2.0\ORADATA\ORCL\USERS01.DBF
SYSAUX     304087040 YES  E:\ORACLE\PRODUCT\10.2.0\ORADATA\ORCL\SYSAUX01.DBF
UNDOTBS1    73400320 YES  E:\ORACLE\PRODUCT\10.2.0\ORADATA\ORCL\UNDOTBS01.DBF
SYSTEM     524288000 YES  E:\ORACLE\PRODUCT\10.2.0\ORADATA\ORCL\SYSTEM01.DBF
EXAMPLE    104857600 YES  E:\ORACLE\PRODUCT\10.2.0\ORADATA\ORCL\EXAMPLE01.DBF
USER01       2097152 NO   E:\ORACLE\PRODUCT\10.2.0\ORADATA\ORCL\USER01.DBF
TEST          761856 YES  E:\ORACLE\PRODUCT\10.2.0\ORADATA\ORCL\TEST.DBF
DLTEST        106496 YES  E:\ORACLE\PRODUCT\10.2.0\ORADATA\ORCL\DLTEST_01.DBF
DLTEST        106496 NO   E:\ORACLE\PRODUCT\10.2.0\ORADATA\ORCL\DLTEST_02.DBF
UNDOTBS2     2097152 YES  E:\ORACLE\PRODUCT\10.2.0\ORADATA\ORCL\UNDO02.DBF
```

已选择 10 行。

查询结果表明,表空间 dltest、undotbs2 都已建成,dltemp 临时表空间在数据字典 dba_data_files 中看不到。在操作系统目录下,同样可以看到表空间对应的数据文件,如图 2-2 所示。

设置系统默认临时表空间为 dltemp,如果不设置,Oracle 将默认 system 表空间为临时表空间。

SQL> alter database default temporary tablespace dltemp;
数据库已更改。

改变表空间的设置,将 dltest_02.dbf 的大小改为 200KB,自动扩展,每次扩展的大小是 100KB,最大可增到 2MB。

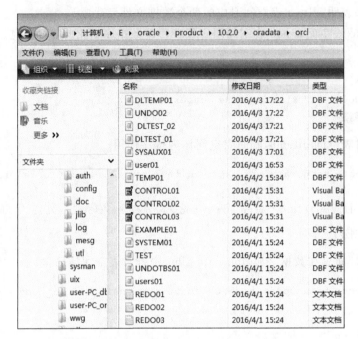

图 2-2　操作系统目录下表空间对应的数据文件

```
SQL> alter database datafile 'e:\Oracle\product\10.2.0\oradata\orcl\ dltest_
02.d
bf' resize 200k;
数据库已更改。
```

将 dltest_02.dbf 改为自动扩展，每次扩展的大小是 100KB，最大可增到 2MB。

```
SQL> alter database datafile 'e:\Oracle\product\10.2.0\oradata\orcl\ dltest_
02.d
bf' autoextend on next 100k maxsize 2m;
数据库已更改。
```

使用 SQL 命令在数据字典 dba_data_files 中查看上述操作。

```
SQL> select tablespace_name,bytes,autoextensible,file_name from dba_data_
files;
TABLESPACE   BYTES AUTO FILE_NAME
----------   ------- ---- --------------------------------------------------
USERS        5242880 YES  E:\ORACLE\PRODUCT\10.2.0\ORADATA\ORCL\USERS01.DBF
SYSAUX       304087040 YES E:\ORACLE\PRODUCT\10.2.0\ORADATA\ORCL\SYSAUX01.DBF
UNDOTBS1     73400320 YES E:\ORACLE\PRODUCT\10.2.0\ORADATA\ORCL\UNDOTBS01.DBF
SYSTEM       524288000 YES E:\ORACLE\PRODUCT\10.2.0\ORADATA\ORCL\SYSTEM01.DBF
EXAMPLE      104857600 YES E:\ORACLE\PRODUCT\10.2.0\ORADATA\ORCL\EXAMPLE01.DBF
USER01       2097152 NO   E:\ORACLE\PRODUCT\10.2.0\ORADATA\ORCL\USER01.DBF
TEST         761856 YES   E:\ORACLE\PRODUCT\10.2.0\ORADATA\ORCL\TEST.DBF
DLTEST       106496 YES   E:\ORACLE\PRODUCT\10.2.0\ORADATA\ORCL\DLTEST_01.DBF
```

```
DLTEST        204800 YES  E:\ORACLE\PRODUCT\10.2.0\ORADATA\ORCL\DLTEST_02.DBF
UNDOTBS2     2097152 YES  E:\ORACLE\PRODUCT\10.2.0\ORADATA\ORCL\UNDO02.DBF
```

已选择 10 行。

查询结果表明，dltest_02.dbf 的大小改为 200KB，autoextend 属性改为 on，临时表空间对应的数据文件不在数据字典 dba_data_files 中。

使用 SQL 命令在数据字典 dba_tablespaces 中查询 CONTENTS 的设置。

```
SQL> select tablespace_name, status,contents from dba_tablespaces;

TABLESPACE STATUS   CONTENTS
---------- -------- ---------
SYSTEM     ONLINE   PERMANENT
UNDOTBS1   ONLINE   UNDO
SYSAUX     ONLINE   PERMANENT
TEMP       ONLINE   TEMPORARY
USERS      ONLINE   PERMANENT
UNDOTBS2   ONLINE   UNDO
EXAMPLE    ONLINE   PERMANENT
USER01     ONLINE   PERMANENT
TEST       ONLINE   PERMANENT
DLTEST     ONLINE   PERMANENT
DLTEMP     ONLINE   TEMPORARY
```

已选择 11 行。

修改表空间，使表空间脱机。进行数据库维护时，需要将需要维护的表空间脱机。将 user01 表空间脱机。

```
SQL> alter tablespace user01 offline normal;
表空间已更改。
```

使用 SQL 命令在数据字典 dba_tablespaces 中查询表空间的状态 online_status。

```
SQL> select tablespace_name,online_status from dba_data_files;

TABLESPACE_NAME                ONLINE_
------------------------------ -------
USERS                          ONLINE
SYSAUX                         ONLINE
UNDOTBS1                       ONLINE
SYSTEM                         SYSTEM
EXAMPLE                        ONLINE
USER01                         OFFLINE
TEST                           ONLINE
```

```
DLTEST                          ONLINE
DLTEST                          ONLINE
UNDOTBS2                        ONLINE
```
已选择 10 行。

查询结果表明，表空间 user01 已脱机，现将表空间 user01 联机。

```
SQL> alter tablespace user01 online;
```
表空间已更改。

使用 SQL 命令在数据字典 dba_tablespaces 中查询表空间的状态 online_status。

```
SQL> select tablespace_name,online_status from dba_data_files;

TABLESPACE_NAME                 ONLINE_
------------------------------  -------
USERS                           ONLINE
SYSAUX                          ONLINE
UNDOTBS1                        ONLINE
SYSTEM                          SYSTEM
EXAMPLE                         ONLINE
USER01                          ONLINE
TEST                            ONLINE
DLTEST                          ONLINE
DLTEST                          ONLINE
UNDOTBS2                        ONLINE
```
已选择 10 行。

删除表空间。使用 drop tablespace 命令删除表空间 dltest、undotbs2、dltemp。使用 contents and datafiles 子句可以同时删除操作系统下的文件。

```
SQL> drop tablespace dltest including contents and datafiles;
```

表空间已删除。
```
SQL> drop tablespace undotbs2 including contents and datafiles;
```
表空间已删除。

```
SQL> drop tablespace dltemp including contents and datafiles;
```
表空间已删除。

使用 SQL 命令在数据字典 dba_data_files 中查看上述操作。

```
SQL> col tablespace_name for a10
SQL> col file_name for a55
SQL> col autoextensible for a4
SQL> set linesize 500
SQL> select tablespace_name,bytes,autoextensible,file_name from dba_data_
```

```
files;

TABLESPACE   BYTES    AUTO FILE_NAME
----------   -------- ---- --------------------------------------------------
USERS        5242880   YES E:\ORACLE\PRODUCT\10.2.0\ORADATA\ORCL\USERS01.DBF
SYSAUX       304087040 YES E:\ORACLE\PRODUCT\10.2.0\ORADATA\ORCL\SYSAUX01.DBF
UNDOTBS1     73400320  YES E:\ORACLE\PRODUCT\10.2.0\ORADATA\ORCL\UNDOTBS01.DBF
SYSTEM       524288000 YES E:\ORACLE\PRODUCT\10.2.0\ORADATA\ORCL\SYSTEM01.DBF
EXAMPLE      104857600 YES E:\ORACLE\PRODUCT\10.2.0\ORADATA\ORCL\EXAMPLE01.DBF
USER01       2097152   NO  E:\ORACLE\PRODUCT\10.2.0\ORADATA\ORCL\USER01.DBF
TEST         761856    YES E:\ORACLE\PRODUCT\10.2.0\ORADATA\ORCL\TEST.DBF
```

已选择 7 行。

查询结果表明，表空间 dltest、undotbs2、dltemp 都已被删除。在操作系统目录下，同样看到表空间 dltest、undotbs2、dltemp 对应的数据文件都已被删除，如图 2-3 所示。

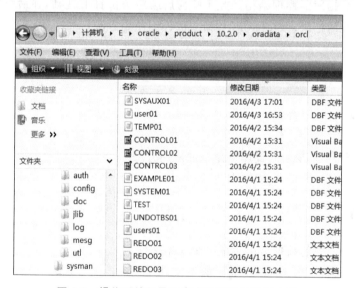

图 2-3　操作系统目录下表空间对应的数据文件

第 3 章

PL/SQL 程序设计

SQL*Plus 命令执行完后,不保存在 SQLBuffer 的内存区域中,它们一般用来对输出的结果进行格式化显示,以便于查看等。用户可以根据需要将环境参数设置成自己所需要的值。本实验中,同学们要认真体会环境变量的使用目的,在生成冷备份脚本的实验中还要分析数据库的物理结构。此实验为后续冷备份打下基础,在后续的冷备份实验中将使用本实验的结果。SQL*Plus 命令的详细使用方法可参考附录。

实验 3.1 SQL*Plus 命令的使用

【实验目的】

(1) 掌握 SQL*Plus 设置环境命令的使用方法。
(2) 使用 SQL*Plus 生成脚本文件。

【实验内容】

(1) 将一个表中的数据导出生成一个文本文件。
(2) 使用 SQL*Plus 生成冷备份脚本文件。

【实验步骤】

(1) 将 emp 表中的数据导出生成一个文本文件,列与列之间以","隔开。

源程序保存为名为 emp 的文本文件,以 sql 作为扩展名,emp.sql 文件的内容如图 3-1 所示。在"另存为"对话框的"保存类型"下拉列表中选择"所有文件"选项,如图 3-2 所示。新建一个文本文件 empout,存放执行 emp.sql 的结果。

在 SQL*Plus 中执行 emp.sql 文件。

```
SQL> @e:\emp.sql
SQL>
```

在生成的 empout.txt 文件中查看结果,如图 3-3 所示。

```
set echo off
set feedback off
set newpage none
set pagesize 5000
set linesize 500
set verify off
set term off
set trims on
set heading off
set timing off
set serveroutput on

SPOOL e:\empout.txt
   select empno || ',' || ename from emp;
SPOOL OFF;
```

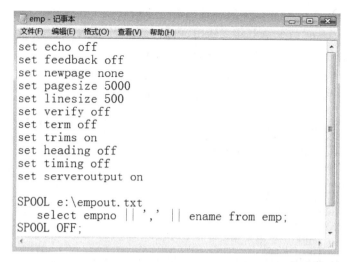

图 3-1 emp.sql 文件的内容

图 3-2 emp.sql 保存类型选择界面

```
0001      ,张蓓
0003      ,黄欣懿
0004      ,邓瑞峰
0005      ,李敏
0006      ,刘思佳
```

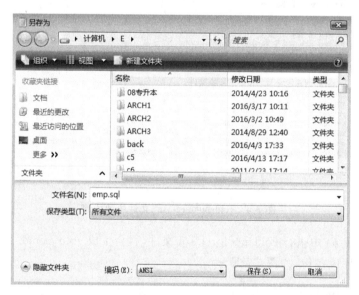

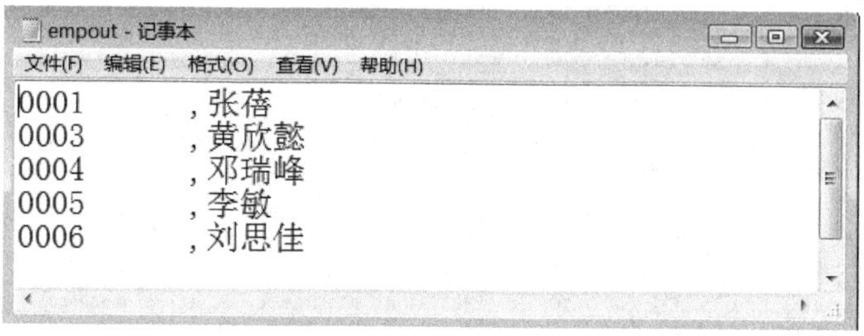

图 3-3 empout.txt 文件内容

(2) 使用 SQL * Plus 动态生成批量脚本。

将 spool 与 select 命令结合起来使用,可以生成一个脚本,脚本中包含有可以批量执行某一任务的语句。建立存放源代码的文件 coldspoolorcl.sql,在 SQL * Plus 中执行 coldspoolorcl.sql 文件,将生成 cold_com.sql 文件,cold_comd.sql 文件中生成了数据库冷备份需要的若干命令。coldspoolorcl.sql 文件中的内容如图 3-4 所示。

图 3-4　coldspoolorcl.sql 文件中的内容

在 SQL * Plus 中执行 coldspoolorcl.sql 文件,就会生成 cold_comd.sql 文件,该文件可以实现数据库冷备份,如图 3-5 所示。

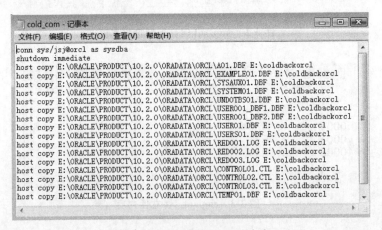

图 3-5　cold_comd.sql 文件内容

```
SQL> @e:\coldspoolorcl.sql
SQL>
```

实验 3.2 PL/SQL 基本编程方法

【实验目的】

(1) 掌握 PL/SQL 块的基本结构、功能及编程规范。
(2) 掌握 PL/SQL 块中各种 SQL 命令的使用方法。
(3) 掌握 PL/SQL 块中流程控制语句的使用方法。
(4) 掌握 PL/SQL 块中游标的使用方法。

【实验内容】

(1) PL/SQL 块的基本结构及使用方法。
(2) PL/SQL 块中流程控制语句的使用方法。
(3) PL/SQL 块中游标的使用方法。
(4) 使用 PL/SQL 生成热备份脚本文件。

【实验步骤】

此实验为后续联机备份做准备,在后续的联机备份实验中将使用本实验的结果。创建一个 hotspool.sql 文件,此文件通过 SQL * Plus 环境变量的设置、spool 命令的使用,运用 PL/SQL 的游标生成热备份脚本文件。以下内容存入 hotspool.sql 文件中。

```
Rem 设置 SQL * Plus 环境参数

set echo off
set feedback off
set pagesize 0
set heading off
set verify off
set linesize 100
set trimspool on

Rem 设置需要使用的 SQL * Plus 用户变量

define dir='e:\hotbackorcl'
define fil='e:\hotbackorcl\hot_com.sql'
define spo='&dir\open_backup_output.lst'
define cpy='copy'
```

```
prompt *** Spooling to &fil

Rem 产生备份命令脚本文件

set serveroutput on
spool &fil
prompt conn sys/jsj@orcl as sysdba

DECLARE
  CURSOR cur_tablespace IS
    SELECT tablespace_name
      FROM dba_tablespaces;
  CURSOR cur_datafile (tn VARCHAR) IS
    SELECT file_name
      FROM dba_data_files
      WHERE tablespace_name=tn;
BEGIN
  FOR ct IN cur_tablespace LOOP
    IF ct.tablespace_name! ='TEMP' then
    dbms_output.put_line ('alter tablespace '||ct.tablespace_name||' begin
    backup;');
    FOR cd IN cur_datafile (ct.tablespace_name) LOOP
    dbms_output.put_line ('host &cpy '||cd.file_name||' &dir');
    END LOOP;
    dbms_output.put_line ('alter tablespace '||ct.tablespace_name||' end
    backup;');
    end if;
  END LOOP;
END;
/

prompt alter system switch logfile;;
prompt alter database backup controlfile to '&dir\backcontrol.ctl' REUSE;;
prompt archive log list;;
spool off;
```

在 SQL*Plus 中执行 hotspool.sql 文件,将生成热备份脚本文件 hot_com.sql,如图 3-6 所示。

```
SQL> @e:\hotspool.sql
SQL>
```

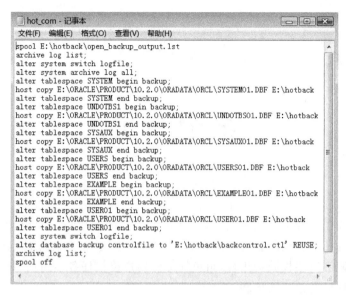

图 3-6 热备份脚本文件 hot_com.sql 内容

第 4 章

安全管理

Oracle 定义了一套丰富完整的权限,通过授予和撤销用户的权限,实现安全的数据库访问控制。为了简化权限管理并提高效率,Oracle 使用角色的概念。角色是具有名称的一组系统权限和对象权限的集合,角色管理使用户授权任务变得简单。以角色为中介将权限和用户账号有效灵活地联系起来。

实验 4.1 用户、权限管理

【实验目的】

(1) 掌握 Oracle 数据库用户的创建与维护。
(2) 掌握 Oracle 数据库权限的分配与管理。
(3) 掌握 Oracle 数据库角色的创建与管理。
(4) 掌握 Oracle 资源配置文件 profile 的创建与管理。

【实验内容】

(1) 用户的建立、查看、修改、删除操作。
(2) 权限的建立、查看、修改、删除操作。
(3) 角色的建立、查看、修改、删除操作。
(4) 资源配置文件 profil 的建立、查看、修改、删除操作。

【实验步骤】

1. 用户管理

(1) 创建用户。

创建用户 user01,密码为 a,默认表空间为 users,临时表空间为 temp,users 表空间使用配额 500MB。

```
SQL> create user user01
```

```
2  identified by a
3  default tablespace users
4  temporary tablespace temp
5  quota 500m on users;
```

用户已创建。

(2) 查看用户。
从 DBA_USERS 数据字典中查看 user01 的信息。

```
SQL> select username, default_tablespace from dba_users where uasername=
'USER01';
USERNAME                        DEFAULT_TABLESPACE
------------------              ------------------------------
USER01                          USERS
```

(3) 修改用户。
修改用户 user01，密码为 b，默认表空间为 user001，表空间 user001 需先建好，用户 user01 使用表空间 user001 的配额为 200MB。

```
SQL> alter user user01
2  identified by b
3  default tablespace user001
4  quota 200m on user001;
```

用户已更改。

(4) 删除用户。
删除用户 user01，带有 cascade 选项，同时删除 user01 用户的所有数据对象。

```
SQL> drop user user01 cascade;
```

用户已删除。

2．权限管理

(1) 授予用户 user01 的 create any table、select any table、create session 系统权限，并用 with admin option 传递权限。

```
SQL> grant create session,create any table, select any table to user01 with
admin option;
```

授权成功。

(2) 从 user_sys_privs 数据库字典中查看用户 user01 的系统权限。

```
SQL> conn user01/b@orcl
```

已连接。
SQL> select username, privilege,admin_option from user_sys_privs;

USERNAME	PRIVILEGE	ADM
USER01	CREATE ANY TABLE	YES
USER01	CREATE SESSION	YES
USER01	SELECT ANY TABLE	YES

（3）验证用户 user01 所拥有的系统权限。

用户 user01 可以成功建表 test，当建序列 tno 时系统报错，因为用户 user01 没有建序列的权限。

```
SQL> conn user01/b@orcl
已连接。
SQL> create table test(a int) tablespace users;
表已创建。

SQL> create sequence tno
2   start with 1
3   increment by 1
4   nomaxvalue;
create sequence tno
 *
第 1 行出现错误：
ORA-01031: 权限不足
```

（4）收回用户 user01 的 create any table、select any table、create session 系统权限。

```
SQL> conn sys/jsj@orcl as sysdba
已连接。
SQL> revoke create session,create any table, select any table from user01;
```

撤销成功。

（5）授予用户 user01 对 sys.student 表的 delete、select 的权限及对 sname 字段 update 的权限，并用 with grant option 传递权限。

```
SQL> grant delete, select, update(sname) on student to user01 with grant option;
```

授权成功。

（6）从 user_tab_privs 数据库字典中查看用户 user01 的对象权限；从 user_col_privs 数据库字典中查看用户 user01 在所有的表列上的对象权限信息。

```
SQL> Select grantee,owner,table_name object_name,privilege,grantable from
```

```
user_tab_privs where grantee='USER01';

GRANTEE     OWNER    OBJECT_NAM   PRIVILEGE   GRA
----------  -------  -----------  ----------  ---
USER01      SYS      STUDENT      DELETE      YES
USER01      SYS      STUDENT      SELECT      YES

SQL> select * from user_col_privs;

GRANTEE   OWNER   TABLE_NAME   COLUMN_NAME   GRANTOR   PRIVILEGE   GRA
--------  ------  -----------  ------------  --------  ----------  ---
USER01    SYS     STUDENT      SNAME         SYS       UPDATE      YES
```

（7）验证用户 user01 所得的对象权限。用户 user01 可以成功 delete 和 select 学生表，当执行 inster 学生表时系统报错，因为用户 user01 没有 insert 的权限。

```
SQL> conn user01/b@orcl
已连接。
SQL> select * from sys.student;

SNO      SNAME            SS   SAGE   SDEPT
-------  ---------------  --   -----  ----------
95020    陈冬              男   18     IS
95021    王红              女   19     IS
95022    张亮              男   19     CS
95023    周明              男   18     IS

SQL> insert into sys.student values ('95004','陈晓','男',21,'cs');
insert into sys.student values ('95004','陈晓','男',21,'cs')
            *
第 1 行出现错误:
ORA-01031: 权限不足
```

（8）收回用户 user01 对 sys.student 表的 delete、select 的权限及对 sname 字段 update 的权限。

```
SQL> conn sys/jsj@orcl as sysdba
已连接。
SQL> revoke delete, select, update on sys.student from user01;

撤销成功。
```

3. 角色管理

（1）创建角色。

创建角色 a，口令为 past。

```
SQL> create role a identified by past;
```

角色已创建。

(2) 管理角色。

赋予角色 a 建立表、索引、视图的权限，并将角色 a 授予用户 user01。用户拥有所得角色的全部权限，可以通过实验验证。在添加或收回角色的某个权限时，用户也同时增加或失去了该权限。

```
SQL> grant create table,create view,create any index to a;
```
授权成功。

```
SQL> grant a to user01;
```
授权成功。

(3) 查看角色。

当前用户拥有角色。

```
SQL> Conn user01/b@orcl;
```
已连接。
```
SQL> select * from session_roles;

ROLE
------------------------------
A
```

(4) 删除角色。

```
SQL> conn sys/jsj@orcl as sysdba
```
已连接。
```
SQL> drop role a;
```

角色已删除。

4. 资源配置文件 PROFILE 的建立、修改、查看、删除操作

(1) 创建 PROFILE 文件。

创建概要文件 user_pro，要求在此概要文件中可以建立 8 个并发的会话连接，登录失败次数为 3，锁定天数为 1，密码有效期为 30 天，宽限时间为 2 天。

```
SQL> create profile user_pro limit
  2  sessions_per_user 8
  3  failed_login_attempts 3
  4  password_lock_time 1/1440
  5  password_life_time 30
  6  password_grace_time 2;
```

配置文件已创建。

（2）查看 PROFILE 文件。

从 DBA_PROFILES 数据字典中查看 user_pro 概要文件的资源名称和资源值等信息。

```
SQL> Select resource_name,resource_type,limit from dba_profiles where
profile='USER_PRO';

RESOURCE_NAME                  RESOURCE  LIMIT
------------------------------ --------  ----------------
COMPOSITE_LIMIT                KERNEL    DEFAULT
SESSIONS_PER_USER              KERNEL    8
CPU_PER_SESSION                KERNEL    DEFAULT
CPU_PER_CALL                   KERNEL    DEFAULT
LOGICAL_READS_PER_SESSION      KERNEL    DEFAULT
LOGICAL_READS_PER_CALL         KERNEL    DEFAULT
IDLE_TIME                      KERNEL    DEFAULT
CONNECT_TIME                   KERNEL    DEFAULT
PRIVATE_SGA                    KERNEL    DEFAULT
FAILED_LOGIN_ATTEMPTS          PASSWORD  3
PASSWORD_LIFE_TIME             PASSWORD  30
PASSWORD_REUSE_TIME            PASSWORD  DEFAULT
PASSWORD_REUSE_MAX             PASSWORD  DEFAULT
PASSWORD_VERIFY_FUNCTION       PASSWORD  DEFAULT
PASSWORD_LOCK_TIME             PASSWORD  .0006
PASSWORD_GRACE_TIME            PASSWORD  2
```

已选择 16 行。

（3）修改 PROFILE 文件。

修改概要文件 user_pro，将登录失败次数改为 2。

```
SQL> alter profile user_pro limit
  2  failed_login_attempts 2;
```

配置文件已更改。

（4）使用 PROFILE 文件。

将概要文件 user_pro 赋予用户 user01，用户 user01 将受到 user_pro 的限制。验证 user01 的登录失败次数，登录两次失败后用户 user01 被锁定。

```
SQL> alter user user01
  2  profile user_pro;
```

用户已更改。

```
SQL> conn user01/z@orcl
ERROR:
ORA-01017: invalid username/password; logon denied


警告：您不再连接到 ORACLE。
SQL> conn user01/x@orcl
ERROR:
ORA-01017: invalid username/password; logon denied


SQL> conn user01/c@orcl
ERROR:
ORA-28000: the account is locked
```

（5）删除 PROFILE 文件。

```
SQL> conn sys/jsj@orcl as sysdba
已连接。
SQL> drop profile user_pro cascade;

配置文件已删除。
```

实验 4.2 触 发 器

【实验目的】

(1) 了解触发器的功能。
(2) 掌握触发器的使用方法。

【实验内容】

(1) 建立触发器。
(2) 验证触发器的结果。
(3) 维护触发器。

【实验步骤】

触发器不能建在 sys 的数据对象上，本例在用户 user01 的数据对象上建立触发器，授予用户 user01 建立触发器的权限。

```
SQL> conn sys/jsj@orcl as sysdba
已连接。
SQL> grant create trigger to user01;
授权成功。
```

切换到 user01 用户，建立 emp 表、dept 表，存放被删除数据的表 emp_his，并插入 emp 表、dept 表的相关数据。

```
SQL> conn user01/a@orcl
已连接。

SQL> create table dept
  2  (deptno number(14) not null,
  3   dname char(20) not null,
  4   loc char(20),
  5   primary key (deptno));
表已创建。

SQL> create table emp
  2  (empno char(10) not null,
  3   ename char(20) not null,
  4   sal smallint,
  5   comm smallint,
  6   job char(20),
  7   hiredate date,
  8   deptno number(14),
  9   primary key (empno),
 10   foreign key(deptno) references dept(deptno));
表已创建。

SQL> insert into dept values (10,'销售','大连');
已创建 1 行。
sql> insert into dept values (20,'采购','大连');
已创建 1 行。
sql> insert into dept values (30,'行政','沈阳');
已创建 1 行。
sql> insert into dept values (40,'人事','大连');
已创建 1 行。
sql> insert into dept values (50,'售后','沈阳');
已创建 1 行。

SQL> insert into emp values ('0001','张蓓',1800,500,'经理',sysdate,'10');
已创建 1 行。
sql> insert into emp values ('0003','黄欣懿',2800,500,'职员',sysdate,'20');
已创建 1 行。
sql> insert into emp values ('0004','邓瑞峰',3800,500,'职员',sysdate,'30');
已创建 1 行。
sql> insert into emp values ('0005','李敏',800,500,'职员',sysdate,'40');
已创建 1 行。
```

1. 高级审计

建立一个 before 触发器,当职工表 emp 表被删除一条记录时,把被删除记录写到职工表的删除日志表 emp_his 中去。

首先建立 emp_his 表。下列 SQL 语句建立了一个和 emp 表结构完全一样的表,只是表中无数据,读者可以自己分析一下。

```
SQL> create table emp_his as select * from emp where 1=2;

SQL> create or replace trigger del_emp
  2  before delete on emp for each row
  3  Begin
  4      --将修改前数据插入到日志记录表 emp_his,以供监督使用。
  5  insert into emp_his(deptno, empno, ename, job, sal, comm, hiredate)
  6  values( :old.deptno, :old.empno, :old.ename, :old.job,:old.sal, :old.comm, :old.hiredate );
  7  End;
  8  /
```

触发器已创建。

验证触发器 del_emp,从 emp 表中删除 empno 为 0001 的记录。

```
SQL> delete emp where empno=0001;
```
已删除 1 行。

执行 SQL 命令,查询 emp 表中数据。

```
SQL> col empno for a5
SQL> col ename for a10
SQL> col hiredat for a10
SQL> col job for a10
SQL> select * from emp;
```

EMPNO	ENAME	SAL	COMM	JOB	HIREDATE	DEPTNO
0003	黄欣懿	2800	500	职员	02-4月-16	20
0004	邓瑞峰	3800	500	职员	02-4月-16	30
0005	李敏	800	500	职员	02-4月-16	40

查询结果表明,0001 号员工已被删除。再查看 emp_his 表,删除的记录是否在此表中。

```
SQL> select * from emp_his;
```

EMPNO	ENAME	SAL	COMM	JOB	HIREDATE	DEPTNO
0001	张蓓	1800	500	经理	02-4月-16	10

查询结果表明,0001 号员工记录已被插入到 emp_his 表中。

2. 保护数据的完整性

利用行触发器实现级联更新。当 dept 表的部门号被修改的时候，emp 表对应的部门号也相应地被修改；当 dept 表的某个部门号被删除的时候，emp 表对应的部门号也同时被删除。利用触发器保护数据的完整性。

```
SQL> create or replace trigger d_update
  2  after delete or update of deptno on dept
  3  for each row
  4  begin
  5  if (updating and :old.deptno ! =:new.deptno)
  6  then update emp
  7  set deptno=:new.deptno
  8  where deptno=:old.deptno;
  9  end if;
 10   if deleting then
 11  delete emp where deptno=:old.deptno;
 12  end if;
 13  end;
 14  /
```

触发器已创建。

验证触发器 d_update 的功能，将 deptno 为 30 的记录改成 100。

```
SQL> update dept set deptno=100 where deptno=30;
```
已更新 1 行。

使用 SQL 命令查看 dept 表和 emp 表是否同时被更新。

```
SQL> select * from dept;
```

DEPTNO	DNAME	LOC
10	销售	大连
20	采购	大连
100	行政	沈阳
40	人事	大连
50	售后	沈阳

已选择 5 行。

```
SQL> select * from emp;
```

EMPNO	ENAME	SAL	COMM	JOB	HIREDATE	DEPTNO
0003	黄欣懿	2800	500	职员	02-4月-16	20
0004	邓瑞峰	3800	500	职员	02-4月-16	100

```
0005        李敏        800        500        职员        02-4月-16            40
```

查询结果表明,dept 表和 emp 表中 deptno 为 30 的记录同时改成了 100。

3. 实现高级约束

限制对 dept 表修改(包括 INSERT、DELETE、UPDATE)的时间范围,即不允许在非工作时间修改 dept 表。

```
SQL> create or replace trigger dept_time
  2  before insert or delete or update
  3  on dept
  4  begin
  5  if (to_char(sysdate,'day') in ('星期六', '星期日')) or (to_char(sysdate,
     'hh24:mi') not between '08:30' and '18:00') then
  6  raise_application_error(-20001,'不是上班时间,不能修改 dept 表');
  7  end if;
  8  end;
  9  /
```

触发器已创建。

验证触发器,向 dept 表插入一条记录。

```
SQL> insert into dept values (60,'人事','上海');
insert into dept values (60,'人事','上海')
            *
第 1 行出现错误:
ORA-20001: 不是上班时间,不能修改 dept 表
ORA-06512: 在 "USER01.DEPT_TIME", line 3
ORA-04088: 触发器 'USER01.DEPT_TIME' 执行过程中出错
```

检验结果表明,触发器限制了在非工作时间修改 dept 表。

4. 触发器的维护

查看触发器的状态。可以查看数据字典 user_triggers,查看当前用户建立的触发器及触发器的状态。

```
SQL> select trigger_name,status from user_triggers;
TRIGGER_NAME                    STATUS
------------------------------  --------
DEPT_TIME                       ENABLED
D_UPDATE                        ENABLED
DEL_EMP                         ENABLED
BIN$BXdZWd6LSZqeFNL514Prgg==$0  ENABLED
```

查询结果表明,实验中建立的三个触发器状态均为启用 ENABLED 状态。

改变触发器的状态,可以使用 alter 命令启用(ENABLED)或禁止(DISABLED)触发器。

```
SQL> alter trigger d_update disable;
```
触发器已更改。

```
SQL> select trigger_name,status from user_triggers;
TRIGGER_NAME                    STATUS
------------------------------  --------
DEPT_TIME                       ENABLED
D_UPDATE                        DISABLED
DEL_EMP                         ENABLED
BIN$BXdZWd6LSZqeFNL514Prgg==$0  ENABLED
```

查询结果表明,触发器 d_update 已经被修改成禁用 DISABLED 状态。

删除触发器。使用 drop 命令删除触发器。

```
SQL> drop trigger d_update;
```
触发器已删除。

实验 4.3　审　　计

审计可以记录对数据库对象的所有操作,即什么时候,什么用户对什么对象进行了什么类型的操作。

审计可以分为语句审计(Statement Auditing)、权限审计(Privilege Auditing)和对象审计(Object Auditing)三类。

【实验目的】

(1) 了解审计的功能。
(2) 掌握启用数据库审计的步骤。
(3) 掌握设置、查看审计的方法。

【实验内容】

(1) 启用数据库审计。
(2) 设置、查看审计。

【实验步骤】

1. 启用数据库审计

通过数据库初始化参数文件中的 AUDIT_TRAIL 初始化参数启用和禁用数据库审

计。AUDIT_TRAIL 的取值如下：
- DB/TRUE：启动审计功能，并且把审计结果存放在数据字典 SYS.AUD$ 表中。
- OS：启动审计功能，并把审计结果存放在操作系统的审计信息中，可以用 AUDIT_FILE_DEST 初始化参数来指定审计文件存储的目录。
- DB_EXTENDED：具有 DB/TRUE 的功能，另外填写 AUD$ 的 SQLBIND 和 SQLTEXT 字段。
- NONE/FALSE：关闭审计功能，系统默认值。

参数 AUDIT_TRAIL 不是动态的，为了使 AUDIT_TRAIL 参数中的改动生效，需关闭数据库并重新启动。查看审计参数：

```
SQL> show parameter audit
NAME                      TYPE         VALUE
------------------------------------------------------------------------------
audit_file_dest           string       E:\ORACLE\PRODUCT\10.2.0\ADMIN\ORCL\ADUMP
audit_sys_operations      boolean      FALSE
audit_trail               string       NONE
```

Audit_sys_Operations：默认为 false，当设置为 true 时，所有 sys 用户（包括以 sysdba、sysoper 身份登录的用户）的操作都会被记录。

查询结果表明，orcl 当前关闭审计功能。激活审计功能，可将 AUDIT_TRAIL=db 写入 orcl 的参数文件 init.ora，重启数据库。

```
SQL> startup pfile=E:\Oracle\product\10.2.0\admin\orcl\pfile\init.ora force;
ORACLE 例程已经启动。
Total System Global Area   612368384 bytes
Fixed Size                   1292036 bytes
Variable Size              171968764 bytes
Database Buffers           432013312 bytes
Redo Buffers                 7094272 bytes
数据库装载完毕。
数据库已经打开。
SQL> show parameter audit
NAME                      TYPE         VALUE
------------------------------------------------------------------------------
audit_file_dest           string       E:\ORACLE\PRODUCT\10.2.0\ADMIN\ORCL\ADUMP
audit_sys_operations      boolean      FALSE
audit_trail               string       DB
```

2. 设置、查看审计

(1) 语句审计(Statement Auditing)。

对预先指定的某些 SQL 语句进行审计。这里从 SQL 语句的角度出发，审计只关心执行的语句。例如，audit TABLE 命令表明对 create table、drop table 语句的执行进行记录，不管该语句是否针对某个对象的操作。

设置语法：

audit sql 语句
  by 用户名
  by session/access
  whenever (not) successful;

- by access：每一个被审计的操作都会生成一条 audit trail。
- by session：一个会话里面同类型的操作只会生成一条 audit trail，默认为 by session。
- whenever successful：操作成功审计。
- whenever not successful：操作不成功审计。省略该子句的话，不管操作成功与否都会审计。

例：对用户 user01 建表、索引的操作进行审计。无论成功与否，每操作一次，记录一次。新建用户 user01，并授予建表、索引的权限。

```
SQL> create user user01
  2  Identified by a
  3  default tablespace users
  4  Temporary tablespace temp
  5  Quota 500k on users;
```

用户已创建。
```
SQL> grant create session,create table to user01;
```
授权成功。

```
SQL> audit table,index by user01 by access;
```
审计已成功。

查看数据字典 DBA_STMT_AUDIT_OPTS（存放 statement 语句级别的审计设置），检查审计设置情况。

```
SQL> col audit_optionl for a10
SQL> col user_name for a10
SQL> select user_name,audit_option,success,failure from DBA_STMT_AUDIT_OPTS;
USER_NAME  AUDIT_OPTION                         SUCCESS     FAILURE
---------- ------------------------------------ ----------  ----------
USER01     TABLE                                BY ACCESS   BY ACCESS
USER01     INDEX                                BY ACCESS   BY ACCESS
```

结果表明，用户 user01 已经成功设置语句审计。对 user01 建表、索引的操作审计，无论成功与否，每操作一次，记录一次。

切换到用户 user01，建立 course 表。

```
SQL> conn user01/a@orcl
已连接。
SQL> create table course
  2   (cno char(4) primary key,
  3    cname char(20) unique,
  4    cpno char(4),
  5    ccredit smallint);
```

表已创建。

为 course 表建立索引，按先行课程号 cpno 升序建唯一索引。

```
SQL> create unique index coucno on course(cpno);
```
索引已创建。

切换到 sys 用户，在数据字典 DBA_AUDIT_OBJECT 中查看审计结果。数据字典 DBA_AUDIT_OBJECT 存放系统中所有对象的审计跟踪记录。

```
SQL> col username for a10
SQL> col timestamp for a10
SQL> col obj_name for a8
SQL> col owner for a6
SQL> col action_name for a15
SQL> select username, to_char(timestamp,'yyyy-mm-dd hh24:mi:ss'),owner,
action_na
me,obj_name from DBA_AUDIT_OBJECT;
USERNAME   TO_CHAR(TIMESTAMP,'  OWNER  ACTION_NAME      OBJ_NAME
---------- -------------------- ------ ---------------- --------
USER01     2016-03-31 10:43:36  USER01 CREATE TABLE     COURSE
USER01     2016-03-31 10:46:45  USER01 CREATE INDEX     CPNO
```

结果表明，用户 user01 的建表和建索引的操作都被记录到数据字典 DBA_AUDIT_OBJECT 中。

如果撤销审计设置，使用命令 noaudit，将已设置的用户 user01 的语句级审计功能撤销。

```
SQL> noaudit table by user01;
审计未成功。
SQL> noaudit index by user01;
审计未成功。
```

如果此时查询数据库字典 DBA_STMT_AUDIT_OPTS，将没有语句审计设置的记录。

```
SQL> select user_name,audit_option,success,failure from DBA_STMT_AUDIT_
```

OPTS;
未选定行。

(2) 权限审计(Privilege Auditing)。

对涉及某些权限的操作进行审计。这里强调"涉及权限",有时候"语句审计"和"权限审计"是相互重复的。

设置语法:

audit sql 权限名称
　　by 用户名
　　by session/access
whenever (not) successful;

例:对用户 usr01 插入表的权限操作进行审计。无论成功与否,每操作一次,记录一次。

SQL> audit insert any table by user01 by access whenever successful;
审计已成功。

查看数据字典 DBA_PRIV_AUDIT_OPTS(存放 privilege 级别的审计设置),检查审计设置情况。

```
SQL> select user_name,privilege,success,failure from DBA_PRIV_AUDIT_OPTS order
by user_name;
USER_NAME  PRIVILEGE                        SUCCESS    FAILURE
---------- -------------------------------- ---------- ----------
USER01     INSERT ANY TABLE                 BY ACCESS  NOT SET
```

结果表明,用户 user01 已经成功设置权限审计,对 user01 插入表的权限操作进行审计。切换到用户 user01,对表 coures 插入数据。

```
SQL> Conn user01/a@orcl
已连接。
SQL> insert into course values ('01','数据库','01',4);
已创建 1 行。
SQL> insert into course values ('02','操作系统','03',3);
已创建 1 行。
```

切换到 sys 用户,在数据字典 DBA_AUDIT_OBJECT 中查看审计结果。

```
SQL> conn sys/jsj@orcl as sysdba
已连接。
SQL> select username, to_char(timestamp,'yyyy-mm-dd hh24:mi:ss'),owner,action_na
me,obj_name from DBA_AUDIT_OBJECT;
USERNAME   TO_CHAR(TIMESTAMP,' OWNER  ACTION_NAME     OBJ_NAME
---------- ------------------- ------ --------------- --------
USER01     2016-03-31 10:43:36 USER01 CREATE TABLE    COURSE
USER01     2016-03-31 10:46:45 USER01 CREATE INDEX    CPNO
USER01     2016-03-31 12:01:28 USER01 INSERT          COURSE
```

结果表明,用户 user01 对表 course 插入数据的操作已被记录到数据字典中。

如果撤销审计设置,使用命令 noaudit,将已设置的用户 user01 的权限级审计功能撤销。

```
SQL> noaudit insert any table by user01 whenever successful;
```
审计未成功。

如果此时查询数据库字典 DBA_PRIV_AUDIT_OPTS,将没有设置插入表权限审计的记录。

```
SQL> select user_name,audit_option,success,failure from DBA_PRIV_AUDIT_
OPTS;
```
未选定行。

(3) 对象审计(Object Auditing)。

审计记录在指定对象上的操作。

设置语法:

```
audit 实体选项 on schema
    by 用户名
    by session/access
    whenever (not) successful;
```

例:对用户 user01 的表 course 插入、删除数据的操作进行审计。对成功的操作,每操作一次,记录一次。

```
SQL> conn sys/jsj@orcl as sysdba
```
已连接。
```
SQL> audit insert,delete on user01.course by access whenever successful;
```
审计已成功。

设置审计时,没有指出具体用户名,系统将对所有用户进行审计。

查看数据字典 DBA_PRIV_AUDIT_OPTS(存放 object 级别的审计设置),检查审计设置情况。

```
SQL> col owner for a10
SQL> col object_name for a10
SQL> col object_type for a10
SQL> select owner,object_name,object_type,del,ins,sel,upd from DBA_OBJ_AUDIT
_OPTS where owner='USER01';
OWNER      OBJECT_NAM OBJECT_TYP DEL   INS   SEL   UPD
---------- ---------- ---------- ----- ----- ----- -----
USER01     COURSE     TABLE      A/-   A/-   -/-   -/-
```

结果表明,已对用户 user01 的表 course 的删除和插入操作设置审计。

新建用户 user02,切换到用户 user02,对表 coures 执行插入、删除数据。

```
SQL> conn sys/jsj@orcl as sysdba
```
已连接。

```
SQL> create user user02 Identified by a;
用户已创建。
SQL> grant create session, select any table, insert any table,delete any table
to user02;
授权成功。
SQL> conn user02/a@orcl
SQL> insert into course values ('03','数字电路','05',2);
已创建 1 行。
SQL> delete from user01.course where cno=01;
已删除 1 行。
SQL> select * from user01.course;
CNO   CNAME                CPNO   CCREDIT
----  -------------------- ----   --------
03    数字电路              05     2
02    操作系统              03     3
```

切换到 SYS 用户,在数据字典 DBA_AUDIT_OBJECT 中查看审计结果。

```
SQL> conn sys/jsj@orcl as sysdba
已连接。
SQL> select username,to_char(timestamp,'yyyy-mm-dd hh24:mi:ss'),owner,action_n
ame,obj_name from DBA_AUDIT_OBJECT;

USERNAME   TO_CHAR(TIMESTAMP,'    OWNER     ACTION_NAME      OBJ_NAME
---------- -------------------    --------  --------------   --------
USER01     2016-03-31 10:43:35    USER01    DROP TABLE       COURSE
USER01     2016-03-31 10:43:36    USER01    CREATE TABLE     COURSE
USER01     2016-03-31 10:46:45    USER01    CREATE INDEX     CPNO
USER01     2016-03-31 12:01:28    USER01    INSERT           COURSE
USER02     2016-03-31 16:16:36    USER01    INSERT           COURSE
USER02     2016-03-31 16:20:25    USER01    DELETE           COURSE
```

结果表明,用户 user02 对 user01 的表 course 插入、删除数据的操作已被记录到数据库的数据字典中。由于未对表 course 的查询操作进行审计,因此用户 user02 对表 course 的查询操作没有被记录到数据字典中。

如果撤销审计设置,使用命令 noaudit,将已设置的用户 user01 的表 course 的对象级审计功能撤销。

```
SQL> noaudit insert,delete on user01.course whenever successful;
审计未成功。
```

所有审计结果都放在表 SYS.AUD$ 中,DBA_AUDIT_OBJECT 是 SYS.AUD$ 的一个视图。如果清除审计信息,便于再存放新的审计信息,可使用下面语句删除审计记录。

```
SQL> delete from sys.aud$;
```

第 5 章 数据库备份与恢复

备份和恢复是为保护数据库免于数据损失,并且在数据损失发生后,重新创建数据的策略和方法。

实验 5.1 冷备份与恢复

【实验目的】

(1) 掌握冷备份的概念。
(2) 掌握冷备份与恢复的方法。

【实验内容】

(1) 使用冷备份脚本文件进行备份。
(2) 使用冷备份脚本文件进行一键备份与恢复。

【实验步骤】

恢复被删除的数据表。恢复数据流程如图 5-1 所示。
生成冷备份脚本。利用实验 3.1 生成的冷备份脚本文件进行 orcl 数据库备份。

SQL> @e:\coldspoolorcl.sql;   //生成的冷备份脚本文件 cold_comd.sql

模拟数据库操作 1,创建数据表 test,插入一条数据 1。

SQL> create table test(a int) tablespace users;
表已创建。
SQL> insert into test values(1);
已创建 1 行。
SQL> commit;
提交完成。
SQL> select * from test;

A

第 5 章 数据库备份与恢复

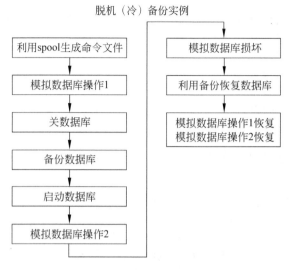

图 5-1 恢复数据流程

                1

利用实验 3.1 生成的冷备份脚本 cold_com.sql 文件备份数据库，cold_comd.sql 文件内容如图 5-2 所示。执行 cold_com.sql 文件，备份数据库 orcl。

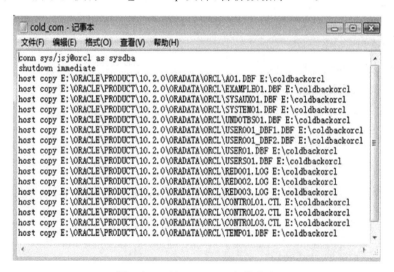

图 5-2 cold_comd.sql 文件内容

```
SQL> @e:\coldbackorcl\cold_com.sql;
已复制        1 个文件。
已复制        1 个文件。
已复制        1 个文件。
已复制        1 个文件。
已复制        1 个文件。
```

已复制   1 个文件。
已复制   1 个文件。
已复制   1 个文件。
已复制   1 个文件。
已复制   1 个文件。
已复制   1 个文件。
已复制   1 个文件。
已复制   1 个文件。

启动数据库，模拟数据库操作 2，插入一条数据 2。

```
SQL> startup
ORACLE 例程已经启动。
Total System Global Area   612368384 bytes
Fixed Size                   1292036 bytes
Variable Size              171968764 bytes
Database Buffers           432013312 bytes
Redo Buffers                 7094272 bytes
数据库装载完毕。
数据库已经打开。

SQL> insert into test values(2);
已创建 1 行。
SQL> commit;
提交完成。
```

查看 test 表，应该有两条记录。

```
SQL> select * from test;
         A
----------
         2
         1
```

模拟数据库损坏，删除表 test。此时查看 test，系统显示表不存在。

```
SQL> drop table test;

SQL> select * from test;
select * from test
       *
第 1 行出现错误：
ORA-00942: 表或视图不存在
```

利用备份进行恢复，因为是冷备份恢复，所以先关闭数据库。

```
SQL> shutdown immediate;
数据库已经关闭。
```

已经卸载数据库。
ORACLE 例程已经关闭。

```
SQL> host copy e:\coldbackorcl\*.* E:\ORACLE\PRODUCT\10.2.0\ORADATA\ORCL\
e:\coldbackorcl\A01.DBF
e:\coldbackorcl\cold_com.sql
e:\coldbackorcl\CONTROL01.CTL
e:\coldbackorcl\CONTROL02.CTL
e:\coldbackorcl\CONTROL03.CTL
e:\coldbackorcl\EXAMPLE01.DBF
e:\coldbackorcl\REDO01.LOG
e:\coldbackorcl\REDO02.LOG
e:\coldbackorcl\REDO03.LOG
e:\coldbackorcl\SYSAUX01.DBF
e:\coldbackorcl\SYSTEM01.DBF
e:\coldbackorcl\TEMP01.DBF
e:\coldbackorcl\UNDOTBS01.DBF
e:\coldbackorcl\USER001_DBF1.DBF
e:\coldbackorcl\USER001_DBF2.DBF
e:\coldbackorcl\USER01.DBF
e:\coldbackorcl\USERS01.DBF
已复制         17 个文件。
```

打开数据库，查看恢复结果。

```
SQL> Startup;
ORACLE 例程已经启动。

Total System Global Area  612368384 bytes
Fixed Size                  1292036 bytes
Variable Size             171968764 bytes
Database Buffers          432013312 bytes
Redo Buffers                7094272 bytes
数据库装载完毕。
数据库已经打开。
SQL> select * from test;

         A
----------
         1
```

查询结果表明，系统只恢复了第一条数据，即备份时间点的数据。备份以后的数据丢失。

恢复被删除的数据文件，并实现使用脚本一键备份及恢复。查看数据库表空间信息。

```
SQL> col tablespace_name for a10
SQL> col file_name for a55
SQL> set linesize 500
SQL> select tablespace_name,file_name from dba_data_files;

TABLESPACE  FILE_NAME
----------  -------------------------------------------------------
USERS       E:\ORACLE\PRODUCT\10.2.0\ORADATA\ORCL\USERS01.DBF
SYSAUX      E:\ORACLE\PRODUCT\10.2.0\ORADATA\ORCL\SYSAUX01.DBF
UNDOTBS1    E:\ORACLE\PRODUCT\10.2.0\ORADATA\ORCL\UNDOTBS01.DBF
SYSTEM      E:\ORACLE\PRODUCT\10.2.0\ORADATA\ORCL\SYSTEM01.DBF
EXAMPLE     E:\ORACLE\PRODUCT\10.2.0\ORADATA\ORCL\EXAMPLE01.DBF
```

新建一个表空间 user01,并在表空间 user01 上建立一个表 test,插入一条数据 1。

```
SQL> create tablespace user01
  2  datafile 'e:\Oracle\product\10.2.0\oradata\orcl\user01.dbf' size
     100k reuse
  3  autoextend on next 100k maxsize 2m;
```

表空间已创建。

```
SQL> create table test(a int) tablespace user01;
```
表已创建。

```
SQL> insert into test values(1);
```
已创建 1 行。

```
SQL> commit;
```
提交完成。

```
SQL> select * from test;
      A
----------
      1
```

在数据字典 user_tables 中查看表 test 所在的表空间。

```
SQL> Select table_name, tablespace_name from user_tables where table_name=
'TEST';

TABLE_NAME                     TABLESPACE
------------------------------ ----------
TEST                           USER01
```

结果表明,表 test 所在的表空间是 USER01。

一键生成备份脚本：@E:\coldspoolorcl.sql；

使用脚本一键备份：@E:\coldbackorcl\cold_com.sql；

模拟数据库文件丢失，删除表空间 user01 对应的数据文件 user01.dbf：

SQL> host del e:\Oracle\product\10.2.0\oradata\orcl\user01.dbf

重新启动数据库，系统将报错，因为数据文件 user01.dbf 已被删除。

```
SQL> startup
ORACLE 例程已经启动。
Total System Global Area    612368384 bytes
Fixed Size                    1292036 bytes
Variable Size               230689020 bytes
Database Buffers            373293056 bytes
Redo Buffers                  7094272 bytes
数据库装载完毕。
ORA-01157: 无法标识/锁定数据文件 6 -请参阅 DBWR 跟踪文件
ORA-01110: 数据文件 6: 'E:\ORACLE\PRODUCT\10.2.0\ORADATA\ORCL\USER01.DBF'
```

使用脚本 coldrestoreorcl.sql 实现一键恢复，脚本 coldrestoreorcl.sql 文件中的内容如图 5-3 所示。

```
SQL> @e:\coldrestoreorcl.sql
E:\coldbackorcl\A01.DBF
E:\coldbackorcl\cold_com.sql
E:\coldbackorcl\CONTROL01.CTL
E:\coldbackorcl\CONTROL02.CTL
E:\coldbackorcl\CONTROL03.CTL
E:\coldbackorcl\EXAMPLE01.DBF
E:\coldbackorcl\REDO01.LOG
E:\coldbackorcl\REDO02.LOG
E:\coldbackorcl\REDO03.LOG
E:\coldbackorcl\SYSAUX01.DBF
E:\coldbackorcl\SYSTEM01.DBF
E:\coldbackorcl\TEMP01.DBF
E:\coldbackorcl\UNDOTBS01.DBF
E:\coldbackorcl\USER001_DBF1.DBF
E:\coldbackorcl\USER001_DBF2.DBF
E:\coldbackorcl\USER01.DBF
E:\coldbackorcl\USERS01.DBF
已复制         17 个文件。
```

打开数据库，查看恢复结果。

```
SQL> startup
ORACLE 例程已经启动。
Total System Global Area               612368384 bytes
```

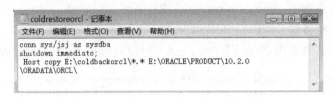

图 5-3　脚本 coldrestoreorcl.sql 文件中的内容

```
Fixed Size                          1292036 bytes
Variable Size                     234883324 bytes
Database Buffers                  369098752 bytes
Redo Buffers                        7094272 bytes
```
数据库装载完毕。
数据库已经打开。

SQL> select * from test;
　　　　　　　　　1

结果表明,数据 1 已被恢复。

实验 5.2　日志文件的管理

Oracle 服务器能够保证所有的已提交数据一定会被记录在重做日志文件上,一旦数据库崩溃,Oracle 服务器就使用重做日志文件中的这些数据进行数据库的恢复工作。可以说引入重做日志文件的目的就是恢复数据库。

【实验目的】

（1）了解日志文件的作用。
（2）掌握重做日志组的添加、删除等操作方法。
（3）掌握重做日志成员的添加、删除等操作方法。

【实验内容】

（1）添加、删除重做日志组。
（2）添加、删除重做日志成员。

【实验步骤】

1. 添加、删除重做日志文件组

在数据库中加入一组新的重做日志组,该组共一个成员,它的文件名为 E:\ORACLE\PRODUCT\10.2.0\ORADATA\ORCL\REDO04.LOG,其大小为 4MB。

使用 SQL 命令查看当前的日志组信息。

```
SQL> select group#,thread#,sequence#,members,archived,status from v$log;

GROUP#    THREAD#   SEQUENCE#  MEMBERS  ARC STATUS
-------   --------- ---------  -------- --- -------
   1         1         20         1     YES INACTIVE
   2         1         21         1     NO  CURRENT
   3         1         19         1     YES INACTIVE
```

查询结果表明，当前数据库共有三个日志组，每组一个日志成员。下面新增一个日志组 GROUP4。

```
SQL> alter database add logfile group 4
E:\ORACLE\PRODUCT\10.2.0\ORADATA\ORCL\REDO04.LOG ') size 4M;
```

数据库已更改。

再使用 SQL 命令验证加入重做日志组命令是否正确。

```
SQL> select group#,thread#,sequence#,members,archived,status from v$log;

GROUP#    THREAD#   SEQUENCE#  MEMBERS  ARC STATUS
-------   --------- ---------  -------- --- -------
   1         1         20         1     YES INACTIVE
   2         1         21         1     NO  CURRENT
   3         1         19         1     YES INACTIVE
   4         1          0         1     YES UNUSED
```

结果表明，日志组 GROUP4 已成功添加。可以使用 SQL 查询语句获取数据库中重做日志组所在的目录及文件名等信息。

```
SQL> select group#,status,type,member from v$logfile;

GROUP#   STATUS    TYPE     MEMBER
------   ------    -----    -----------------------------------------
  3      STALE     ONLINE   E:\ORACLE\PRODUCT\10.2.0\ORADATA\ORCL\REDO03.LOG
  2                ONLINE   E:\ORACLE\PRODUCT\10.2.0\ORADATA\ORCL\REDO02.LOG
  1      STALE     ONLINE   E:\ORACLE\PRODUCT\10.2.0\ORADATA\ORCL\REDO01.LOG
  4                ONLINE   E:\ORACLE\PRODUCT\10.2.0\ORADATA\ORCL\REDO04.LOG
```

当日志组不需要时，可以使用 SQL 命令删除。将刚加的日志组 GROUP4 删除。

```
SQL> alter database drop logfile group 4;
```

数据库已更改。

查看当前日志组信息及日志组所在的目录及文件名信息，验证删除命令。

```
SQL> select group#,thread#,sequence#,members,archived,status from v$log;

   GROUP#    THREAD#    SEQUENCE#    MEMBERS  ARC  STATUS
---------- ---------- ---------- ---------- ----- -------
       1         1          20           1    YES  INACTIVE
       2         1          21           1    NO   CURRENT
       3         1          19           1    YES  INACTIVE
```

查询结果表明，日志组 GROUP4 已被删除。再查看数据字典 v$logfile 信息。

```
SQL> select * from v$logfile;
GROUP#   STATUS   TYPE    MEMBER
-------- -------- ------- ------------------------------------------------
3        STALE    ONLINE  E:\ORACLE\PRODUCT\10.2.0\ORADATA\ORCL\REDO03.LOG
2                 ONLINE  E:\ORACLE\PRODUCT\10.2.0\ORADATA\ORCL\REDO02.LOG
1        STALE    ONLINE  E:\ORACLE\PRODUCT\10.2.0\ORADATA\ORCL\REDO01.LOG
```

查询结果表明，日志组 GROUP4 对应的日志文件 REDO04.LOG 已被删除。

当前的重做日志组不能删除，如果要删除，先使用命令 alter system switch logfile 进行切换。每个实例至少有两组重做日志才能正常工作，当一组重做日志被删除后，它的操作系统文件依然存在，只能使用操作系统命令删除，否则会留下一些无用的垃圾文件。

2. 添加、删除重做日志成员

Oracle 默认安装的是每个重做日志组的一个成员，这对绝大多数生产数据库来说是不安全的。因此在这种情况下，数据库管理员应该在每个重做日志组中再增加至少一个新成员，以防止重做日志文件的物理错误。下面为日志组 GROUP1、GROUP2 各创建一个新的重做日志成员 REDO01B.LOG 和 REDO02B.LOG，文件大小和原来的日志成员一样。为了文件的安全，新建的日志成员放到其他目录上，这里放到 D:\Oracle\orclback 目录下。SQL 命令如下：

```
SQL> alter database add logfile member 'D:\Oracle\orclback\REDO01_2.LOG' to
group 1,D:\Oracle\orclback\REDO02_2.LOG' to group 2;
```

数据库已更改。

再使用 SQL 命令验证加入重做日志成员命令是否正确。

```
SQL> select group#,thread#,sequence#,members,archived,status from v$log;

   GROUP#    THREAD#    SEQUENCE#    MEMBERS  ARC  STATUS
---------- ---------- ---------- ---------- ----- -------
       1         1          23           2    YES  INACTIVE
       2         1          24           2    NO   CURRENT
       3         1          22           1    YES  INACTIVE
```

查询结果表明，日志组 GROUP1、GROUP2 各增加了一个新的重做日志成员。查看当前日志组信息及日志组所在的目录及文件名信息。

```
SQL> select group#,status,type,member from v$logfile;

GROUP#    STATUS    TYPE     MEMBER
------    ------    ------   ----------------------------------------------
3                   ONLINE   E:\ORACLE\PRODUCT\10.2.0\ORADATA\ORCL\REDO03.LOG
2                   ONLINE   E:\ORACLE\PRODUCT\10.2.0\ORADATA\ORCL\REDO02.LOG
1                   ONLINE   E:\ORACLE\PRODUCT\10.2.0\ORADATA\ORCL\REDO01.LOG
1                   ONLINE   D:\ORACLE\ORCLBACK\REDO01_2.LOG
2                   ONLINE   D:\ORACLE\ORCLBACK\REDO02_2.LOG
```

从查询结果可以看出，GROUP1、GROUP2 重做日志组都添加了一个新成员，它们存在 D:\ORACLE\ORCLBACK 目录下，文件名分别为 REDO01_2.LOG 和 REDO02_2.LOG 。

当日志成员不需要时，可以使用 SQL 命令删除。

```
SQL> alter database drop logfile member 'D:\Oracle\orclback\REDO01_2.LOG';
数据库已更改。

SQL> select group#,status,type,member from v$logfile;

GROUP#    STATUS    TYPE     MEMBER
------    ------    ------   ----------------------------------------------
3                   ONLINE   E:\ORACLE\PRODUCT\10.2.0\ORADATA\ORCL\REDO03.LOG
2                   ONLINE   E:\ORACLE\PRODUCT\10.2.0\ORADATA\ORCL\REDO02.LOG
1                   ONLINE   E:\ORACLE\PRODUCT\10.2.0\ORADATA\ORCL\REDO01.LOG
2                   ONLINE   D:\ORACLE\ORCLBACK\REDO02_2.LOG
```

查询结果表明，日志成员 REDO01_2.LOG 已被删除。不能使用上面的命令同时删除每个重做日志组的一个成员，因为不能删除当前组的成员，如果要删除，应该先使用 alter system switch logfile 命令进行切换。如果要删除的日志成员状态为 INVALID，则要执行 alter system switch logfile 命令进行切换，直到日志成员的状态为 INACTIVE，才可以删除。日志成员被删除后，它的操作系统文件仍存在，需要使用操作系统命令删除。

实验 5.3 归档模式的管理

如果数据库运行在非归档模式，即没有产生归档文件的话，数据库无法保证在系统崩溃之后，所有提交的数据都能恢复。在非归档模式下，数据库只能保证恢复到上一次备份的时间点，从上一次备份到系统崩溃这段时间内所有提交的数据会丢失。有了归档日志，数据库就能保证所提交的数据都能恢复，因为那些在重做日志文件中被覆盖的信息已经存在于归档日志文件中了。

【实验目的】

(1) 了解归档文件的作用。
(2) 掌握归档模式的设置。
(3) 掌握归档进程和归档文件目录的管理。

【实验内容】

(1) 将数据库设为归档模式。
(2) 归档进程和归档文件目录的设置。
(3) 归档进程和归档文件目录的管理。

【实验步骤】

1. 将数据库设为归档模式

Oracle 数据库默认是非归档模式，如果想让数据库运行在归档模式，就必须重新设置，将数据库设置为归档模式。

查看数据库的归档模式。

```
SQL> archive log list
数据库日志模式              非存档模式
自动存档                   禁用
存档终点                   USE_DB_RECOVERY_FILE_DEST
最早的联机日志序列           22
当前日志序列                24
```

执行 archive log list 命令，可以看到数据库当前工作在非归档模式，如果使数据库工作在归档模式，首先使用 shutdown immediate 命令关闭数据库。

```
SQL> shutdown immediate
数据库已经关闭。
已经卸载数据库。
ORACLE 例程已经关闭。
```

以 mount 方式启动数据库。

```
SQL> startup mount
ORACLE 例程已经启动。

Total System Global Area    612368384 bytes
Fixed Size                    1292036 bytes
```

```
Variable Size                171968764 bytes
Database Buffers             432013312 bytes
Redo Buffers                   7094272 bytes
```
数据库装载完毕。

使用 alter database archivelog 命令将数据库设为归档模式。

```
SQL> alter database archivelog;
```
数据库已更改。

使用 alter database open 命令打开数据库。

```
SQL> alter database open;
```
数据库已更改。

再用 archive log list 命令验证数据库的归档设置。

```
SQL> archive log list;
```
数据库日志模式	存档模式
自动存档	启用
存档终点	USE_DB_RECOVERY_FILE_DEST
最早的联机日志序列	22
下一个存档日志序列	24
当前日志序列	24

可以看到，数据库已经设为归档模式。自动存档启用表示后台归档写进程启动。

2. 归档进程和归档文件目录的设置

重做日志写进程 LGWR，负责将重做日志缓冲区的信息写到重做日志文件中，而归档进程 ARCn 是把切换后的重做日志文件复制到归档日志文件中。如果数据库操作非常频繁，ARCn 的读写可能跟不上 LGWR，这样可能造成重做日志组已经切换一圈了，归档进程 ARCn 还没有将重做日志文件中的数据归档到归档文件中，日志写进程 LGWR 必须等待重做日志文件中的提交数据被复制的归档文件后，才能写重做日志文件，此时数据库已被挂起。

为了解决以上问题，可以启动多个归档后台进程，以避免由于归档进程 ARCn 跟不上日志写进程 LGWR 而造成数据库系统效率的下降。可以通过修改参数文件 LOG_ARCHIVE_MAX_PROCESSES 的参数来决定启动几个 ARCn 后台进程。LOG_ARCHIVE_MAX_PROCESSES 是动态参数，可以使用 ALTER SYSTEM 命令改变归档进程的个数。

查看当前 ARCn 进程个数。这个参数默认值是 2。

```
SQL> show parameter LOG_ARCHIVE_MAX_PROCESSES

NAME                                 TYPE        VALUE
------------------------------------ ----------- ---------
```

```
log_archive_max_processes              integer     2
```

将 ARCn 进程个数设为 3。

```
SQL> alter system set LOG_ARCHIVE_MAX_PROCESSES=3;
```
系统已更改。

```
SQL> show parameter LOG_ARCHIVE_MAX_PROCESSES
NAME                                   TYPE        VALUE
-----------------------------------    --------    --------
log_archive_max_processes              integer     3
```

归档日志文件存储了数据库恢复所需的所有信息,如果归档日志文件损坏了,数据库的全部恢复是很难做到的,所以对归档日志文件要采取保护措施,以防止由于磁盘和文件的损坏而造成的数据丢失。保护措施就是将多个完全相同的归档日志文件写到不同的物理硬盘上。使用 LOG_ARCHIVE_DEST_n 参数可以定义最多 10 个归档目的地。LOG_ARCHIVE_DEST_n 参数有三个选项:

- MANDATORY:此选项表示必须归档完成后才可以进行日志切换,如果由于介质损坏不能进行日志切换时,数据库处于等待状态。
- REOPEN:该参数表示当归档日志目录设置错误,磁盘介质损坏后,等待时间。
- OPTIONAL:该选项表示不管日志归档是否完成就切换,这是默认方式。

下面设置两个归档日志目录,分别为 E:\ARCH1 属性 MANDATORY 和 E:\ARCH2 属性 OPTIONAL。

查看当前所有的归档日志文件的路径。

```
SQL> show parameter log_archive_dest_

NAME                                   TYPE        VALUE
-----------------------------------    ----------  --------------
log_archive_dest_1                     string
log_archive_dest_10                    string
log_archive_dest_2                     string
log_archive_dest_3                     string
log_archive_dest_4                     string
log_archive_dest_5                     string
log_archive_dest_6                     string
log_archive_dest_7                     string
log_archive_dest_8                     string
log_archive_dest_9                     string
log_archive_dest_state_1               string      enable
log_archive_dest_state_10              string      enable
log_archive_dest_state_2               string      enable
log_archive_dest_state_3               string      enable
log_archive_dest_state_4               string      enable
```

```
log_archive_dest_state_5           string      enable
log_archive_dest_state_6           string      enable
log_archive_dest_state_7           string      enable
log_archive_dest_state_8           string      enable
log_archive_dest_state_9           string      enable
```

显示结果表明，LOG_ARCHIVE_DEST_1～LOG_ARCHIVE_DEST_10 都为空，使用 alter system set LOG_ARCHIVE_DEST_n 设置归档路径。

```
SQL> alter system set log_archive_dest_1="LOCATION=E:\ARCH1 MANDATORY";
系统已更改。

SQL> alter system set log_archive_dest_2="LOCATION=E:\ARCH2";
系统已更改。

SQL> show parameter log_archive_dest_

NAME                                TYPE        VALUE
----------------------------------- ----------- ------------------------------
log_archive_dest_1                  string      LOCATION=E:\ARCH1 MANDATORY
log_archive_dest_10                 string
log_archive_dest_2                  string      LOCATION=E:\ARCH2
log_archive_dest_3                  string
log_archive_dest_4                  string
log_archive_dest_5                  string
log_archive_dest_6                  string
log_archive_dest_7                  string
log_archive_dest_8                  string
log_archive_dest_9                  string
log_archive_dest_state_1            string      enable
log_archive_dest_state_10           string      enable
log_archive_dest_state_2            string      enable
log_archive_dest_state_3            string      enable
log_archive_dest_state_4            string      enable
log_archive_dest_state_5            string      enable
log_archive_dest_state_6            string      enable
log_archive_dest_state_7            string      enable
log_archive_dest_state_8            string      enable
log_archive_dest_state_9            string      enable
```

可以看到，log_archive_dest_1 和 log_archive_dest_2 已设置成功。下面可从数据字典中看到归档路径的属性。

```
col destination for a30
SQL> select destination,binding,status from v$archive_dest;
```

```
DESTINATION                      BINDING    STATUS
-------------------------------  ---------  ---------
E:\ARCH1                         MANDATORY  VALID
E:\ARCH2                         OPTIONAL   VALID
                                 OPTIONAL   INACTIVE
                                 OPTIONAL   INACTIVE
                                 OPTIONAL   INACTIVE
                                 OPTIONAL   INACTIVE
                                 OPTIONAL   INACTIVE
                                 OPTIONAL   INACTIVE
                                 OPTIONAL   INACTIVE
                                 OPTIONAL   INACTIVE
```

这时可以到操作系统的目录下查看归档目录 E:\ARCH1 和 E:\ARCH2 里是否产生了归档日志,如图 5-4 所示。

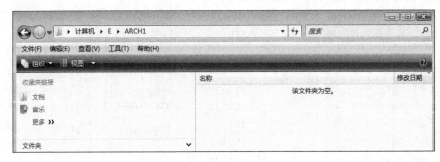

图 5-4 操作系统的归档目录

显示结果表明,归档目录下没有任何归档文件,这是因为数据库没有执行任何的 DML 操作,所以重做日志文件不能被填满,因此不可能产生重做日志切换,也就不可能产生归档日志文件。为了产生归档日志文件,可以使用手动切换命令 alter system switch logfile 强行切换日志。

```
SQL> alter system switch logfile;
系统已更改。
```

再到操作系统下查看,可以看到 E:\ARCH1 和 E:\ARCH2 下产生了镜像的归档日志文件,如图 5-5 和图 5-6 所示。

3. 归档进程和归档文件目录的管理

数据库为了保证归档日志成功,引入了 LOG_ARCHIVE_MIN_SUCCEED_DEST 和 LOG_ARCHIVE_DEST_STATE_n 两个参数。

通过定义参数 LOG_ARCHIVE_MIN_SUCCEED_DEST 的值来限定 Oracle 系统必须保证成功的归档日志文件镜像个数。

首先查看一下 LOG_ARCHIVE_MIN_SUCCEED_DEST 参数的当前值。

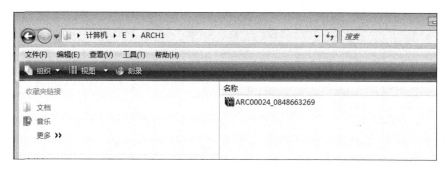

图 5-5　操作系统的归档目录 ARCH1

图 5-6　操作系统的归档目录 ARCH2

```
SQL> show parameter LOG_ARCHIVE_MIN_SUCCEED_DEST;

NAME                                 TYPE        VALUE
------------------------------------ ----------- ---------
log_archive_min_succeed_dest         integer     1
```

LOG_ARCHIVE_MIN_SUCCEED_DEST 参数的当前值为 1，即数据库只要成功写一个物理路径下的归档日志就可以切换日志。现在将 LOG_ARCHIVE_MIN_SUCCEED_DEST 参数的值设为 2。

```
SQL> alter system set LOG_ARCHIVE_MIN_SUCCEED_DEST=2;
系统已更改。

SQL> show parameter LOG_ARCHIVE_MIN_SUCCEED_DEST;

NAME                                 TYPE        VALUE
------------------------------------ ----------- ---------
log_archive_min_succeed_dest         integer     2
```

这样，数据库必须成功写两个物理路径下的归档日志才可以切换日志。

数据库通过定义参数 LOG_ARCHIVE_DEST_STATE_n 的值控制归档的目的地有效（enable）或无效（defer）。

首先查看一下 LOG_ARCHIVE_DEST_STATE_n 参数的当前值。

```
SQL> show parameter log_archive_dest_;

NAME                                  TYPE        VALUE
------------------------------------  ----------  ------------------------------
log_archive_dest_1                    string      LOCATION=E:\ARCH1 MANDATORY
log_archive_dest_10                   string
log_archive_dest_2                    string      LOCATION=E:\ARCH2
log_archive_dest_3                    string
log_archive_dest_4                    string
log_archive_dest_5                    string
log_archive_dest_6                    string
log_archive_dest_7                    string
log_archive_dest_8                    string
log_archive_dest_9                    string
log_archive_dest_state_1              string      enable
log_archive_dest_state_10             string      enable
log_archive_dest_state_2              string      enable
log_archive_dest_state_3              string      enable
log_archive_dest_state_4              string      enable
log_archive_dest_state_5              string      enable
log_archive_dest_state_6              string      enable
log_archive_dest_state_7              string      enable
log_archive_dest_state_8              string      enable
log_archive_dest_state_9              string      enable
```

从显示结果可以看出,所有的归档路径都是 enable,现使用命令使路径 log_archive_dest_1 失效。

```
SQL> alter system set log_archive_dest_state_1 =defer;
系统已更改。

SQL> show parameter log_archive_dest_;

NAME                                  TYPE        VALUE
------------------------------------  ----------  ------------------------------
log_archive_dest_1                    string      LOCATION=E:\ARCH1 MANDATORY
log_archive_dest_10                   string
log_archive_dest_2                    string      LOCATION=E:\ARCH2
log_archive_dest_3                    string
log_archive_dest_4                    string
log_archive_dest_5                    string
log_archive_dest_6                    string
log_archive_dest_7                    string
log_archive_dest_8                    string
log_archive_dest_9                    string
```

```
log_archive_dest_state_1              string      DEFER
log_archive_dest_state_10             string      enable
log_archive_dest_state_2              string      enable
log_archive_dest_state_3              string      enable
log_archive_dest_state_4              string      enable
log_archive_dest_state_5              string      enable
log_archive_dest_state_6              string      enable
log_archive_dest_state_7              string      enable
log_archive_dest_state_8              string      enable
log_archive_dest_state_9              string      enable
```

从显示结果可以看出,只有 log_archive_dest_state_1 的 VALUE 为 DEFER,其他归档路径的 VALUE 都是 enable。现在重新开启 log_archive_dest_state_1 路径。

```
SQL> alter system set log_archive_dest_state_1=enable;
系统已更改。

SQL> show parameter log_archive_dest_;

NAME                                  TYPE        VALUE
------------------------------------- ----------- -------------------------------
log_archive_dest_1                    string      LOCATION=E:\ARCH1 MANDATORY
log_archive_dest_10                   string
log_archive_dest_2                    string      LOCATION=E:\ARCH2
log_archive_dest_3                    string
log_archive_dest_4                    string
log_archive_dest_5                    string
log_archive_dest_6                    string
log_archive_dest_7                    string
log_archive_dest_8                    string
log_archive_dest_9                    string
log_archive_dest_state_1              string      ENABLE
log_archive_dest_state_10             string      enable
log_archive_dest_state_2              string      enable
log_archive_dest_state_3              string      enable
log_archive_dest_state_4              string      enable
log_archive_dest_state_5              string      enable
log_archive_dest_state_6              string      enable
log_archive_dest_state_7              string      enable
log_archive_dest_state_8              string      enable
log_archive_dest_state_9              string      enable
```

显示结果表明,所有的归档路径又都是可用的 enable。defer 状态是一种临时的维护状态,一旦维护工作结束后,如硬盘已经修好,就要及时转换为 enable 状态。当归档日志的物理路径状态被设置为 defer 时,Oracle 不会对这个路径进行归档操作,如果以后将该路径的状态改回为 enable，所有丢失的归档文件必须手工恢复。

实验 5.4 联 机 备 份

数据库运行在归档模式，Oracle 不但可以进行联机备份，而且还可以进行表空间及数据文件一级的联机备份。在进行联机备份时，不用关闭数据库，所有数据库操作可以照常进行。

【实验目的】

（1）了解联机备份的优缺点。
（2）掌握联机备份的步骤。
（3）掌握自动联机备份的方法。

【实验内容】

（1）联机备份的步骤。
（2）自动联机备份的方法。

【实验步骤】

1. 备份单个表空间

联机备份单个表空间的基本步骤如下：
（1）使用数据字典 dba_data_files 查看数据文件及对应的表空间相关信息。

select file_id,file_name,tablespace_name from dba_data_files;

（2）使用数据字典 v$backup 查看当前备份状态。

select * from v$backup;

（3）设置表空间为备份状态。

alter tablespace <表空间> begin backup;

（4）备份表空间的数据文件。
使用操作系统命令 copy 复制表空间文件到备份位置。
（5）结束表空间备份状态。

alter tablespace <表空间> end backup;

（6）将当前联机日志文件写到归档日志文件中。
备份单个表空间 users。首先从数据字典 dba_data_files 中查询数据文件及对应的表空间的相关信息。

```
SQL> col file_name for a55
SQL> col tablespace_name for a10
SQL> select file_id,file_name,tablespace_name from dba_data_files;

   FILE_ID FILE_NAME                                               TABLESPACE
---------- ------------------------------------------------------- ----------
         4 E:\ORACLE\PRODUCT\10.2.0\ORADATA\ORCL\USERS01.DBF       USERS
         3 E:\ORACLE\PRODUCT\10.2.0\ORADATA\ORCL\SYSAUX01.DBF      SYSAUX
         2 E:\ORACLE\PRODUCT\10.2.0\ORADATA\ORCL\UNDOTBS01.DBF     UNDOTBS1
         1 E:\ORACLE\PRODUCT\10.2.0\ORADATA\ORCL\SYSTEM01.DBF      SYSTEM
         5 E:\ORACLE\PRODUCT\10.2.0\ORADATA\ORCL\EXAMPLE01.DBF     EXAMPLE
         6 E:\ORACLE\PRODUCT\10.2.0\ORADATA\ORCL\USER01.DBF        USER01
```

已选择 6 行。

从数据字典 v$backup 中查询每个数据文件的备份状态信息。

```
SQL> select * from v$backup;

     FILE# STATUS                  CHANGE# TIME
---------- ------------------ ------------ -------------
         1 NOT ACTIVE             23173030 20-5月 -14
         2 NOT ACTIVE             23173047 20-5月 -14
         3 NOT ACTIVE             23173060 20-5月 -14
         4 NOT ACTIVE             23173075 20-5月 -14
         5 NOT ACTIVE             23173093 20-5月 -14
         6 NOT ACTIVE             23556015 12-5月 -15
```

已选择 6 行。

查询结果显示，所有数据文件都处于非备份状态 NOT ACTIVE。将表空间 USERS 设置成备份状态，再查看数据文件的备份状态信息。

```
SQL> alter tablespace users begin backup;
```
表空间已更改。

```
SQL> select * from v$backup;

     FILE# STATUS                  CHANGE# TIME
---------- ------------------ ------------ -------------
         1 NOT ACTIVE             23173030 20-5月 -14
         2 NOT ACTIVE             23173047 20-5月 -14
         3 NOT ACTIVE             23173060 20-5月 -14
         4 ACTIVE                 23852551 01-3月 -16
         5 NOT ACTIVE             23173093 20-5月 -14
         6 NOT ACTIVE             23556015 12-5月 -15
```

已选择 6 行。

从显示结果可以看到,表空间 USERS 对应的 4 号数据文件处在备份状态 ACTIVE。使用操作系统命令复制表空间文件到备份位置 e:\hotback。

SQL> host copy e:\Oracle\product\10.2.0\oradata\orcl\users01.dbf e:\hotback
已复制   1 个文件。

结束表空间 USERS 的备份状态。再查看数据文件的备份状态信息。

SQL> alter tablespace users end backup;
表空间已更改。

SQL> select * from v$backup;

FILE#	STATUS	CHANGE#	TIME
1	NOT ACTIVE	23173030	20-5月-14
2	NOT ACTIVE	23173047	20-5月-14
3	NOT ACTIVE	23173060	20-5月-14
4	NOT ACTIVE	23852551	01-3月-16
5	NOT ACTIVE	23173093	20-5月-14
6	NOT ACTIVE	23556015	12-5月-15

已选择 6 行。

结果表明,4 号数据文件又恢复到非备份状态。再查看一下操作系统文件是否已经生成,如图 5-7 所示。

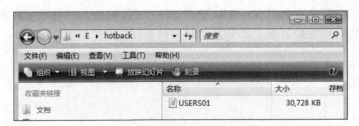

图 5-7　生成的操作系统文件

可以看到操作系统文件已经生成了。

查看当前归档日志信息,使用命令将当前联机日志文件写到归档日志文件中,再查看当前归档日志信息。

SQL> select group#,thread#,sequence#,members,archived,status from v$log;

GROUP#   THREAD#   SEQUENCE#   MEMBERS   ARC   STATUS

```
----------  ----------  ----------  ----------  ---  ----------------
    1           1           26          1       NO   CURRENT
    2           1           24          2       YES  INACTIVE
    3           1           25          1       YES  INACTIVE
```

结果表明,当前的联机日志序列号是 26 号。使用 alter system switch logfile 命令手动切换日志,再查看当前归档日志信息。

```
SQL> alter system switch logfile;
系统已更改。

SQL> select group#,thread#,sequence#,members,archived,status from v$log;

  GROUP#      THREAD#     SEQUENCE#   MEMBERS    ARC  STATUS
----------  ----------  ----------  ----------  ---  ----------------
    1           1           26          1       YES  ACTIVE
    2           1           27          2       NO   CURRENT
    3           1           25          1       YES  INACTIVE
```

结果表明,26 号日志已被归档,当前的联机日志序列号是 27 号。再查看一下操作系统归档文件是否已经生成,如图 5-8 所示。

图 5-8 操作系统中的归档日志文件

显示结果表明,一个序列号为 26 号的归档日志已经生成。

2. 自动联机备份

可以将备份命令写成一个 SQL 脚本文件,备份时执行该脚本文件即可。这样可以减轻 DBA 备份的工作量,并减少出错的几率。可利用实验 3.2 的代码生成一个完全备份脚本,也可以根据需要将备份单个表空间的备份命令写成一个脚本。

执行实验 3.2 的代码生成备份脚本。

```
SQL> @e:\hotspool.sql
```

脚本文件 hotspool.sql 将在 e:\hotback 目录下生成另一个完全备份脚本文件 hot_com.sql。打开 hot_com.sql 查看内容,如图 5-9 所示。

执行 hot_com.sql 脚本,首先将在 E:\hotback 路径下生成一个名为 open_backup_output.lst 的备份提示文件,该文件记录备份前的联机日志信息及备份后的联机日志信

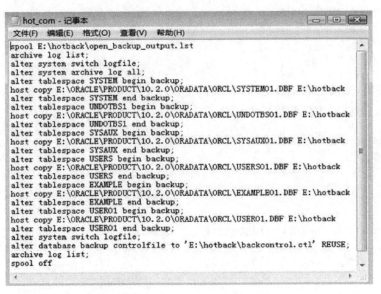

图 5-9 hot_com.sql 文件内容

息,主要目的为 DBA 进行数据库恢复时查询备份时的联机日志序列号。此项操作也可以省略。在脚本中使用 archive log list 命令显示当前联机日志信息,再使用 alter system switch logfile 命令对数据库中的联机日志执行强制切换,归档当前联机日志,然后使用 alter system archive log all 命令对数据库中的非当前未归档日志进行归档,不负责归档 current 日志。如果没有未归档的非当前日志,系统会提示"ORA-00271:没有需要归档的日志。"这两个命令的主要目的是在接下来的数据库备份前使所有的日志文件都得到归档。数据库备份时如果出现错误,不会殃及到日志文件。

数据库共备份了 6 个数据文件,将控制文件备份到 E:\hotback\backcontrol.ctl,这是一个完全数据库备份方案。备份结束后再次使用 alter system switch logfile 命令强行切换日志,使当前日志得到归档。目的是为了在数据库联机备份期间记录对数据库进行的所有的 DML 操作。最后使用 archive log list 命令显示当前联机日志信息。

```
SQL> @e:\hotback\hot_com.sql
数据库日志模式          存档模式
自动存档              启用
存档终点              E:\ARCH2
最早的联机日志序列       29
下一个存档日志序列       31
当前日志序列           31
alter system archive log all
*
第 1 行出现错误:
ORA-00271:没有需要归档的日志


已复制        1 个文件。
```

已复制   1 个文件。

已复制   1 个文件。

已复制   1 个文件。

已复制   1 个文件。

已复制   1 个文件。

数据库日志模式   存档模式
自动存档   启用
存档终点   E:\ARCH2
最早的联机日志序列   31
下一个存档日志序列   33
当前日志序列   33

从显示结果可以推断出，数据库联机备份时使用的 current 日志序列号是 32 号。这些信息已被写到 open_backup_output.lst 的备份提示文件中。在操作系统下可以查看 open_backup_output.lst 文件的内容，如图 5-10 所示。

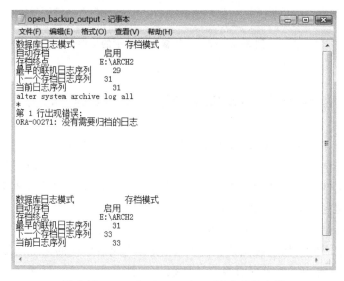

图 5-10 open_backup_output.lst 文件的内容

在操作系统归档路径 e:\arch2 下可以看到被归档的 32 号日志文件，如图 5-11 所示。

在操作系统备份路径 E:\hotback 下可以看到已被备份的 6 个数据文件和 1 个控制文件，如图 5-12 所示。

图 5-11 归档的日志文件

图 5-12 操作系统备份路径下的文件

实验 5.5 归档模式下的数据库恢复

在归档模式下恢复数据库时,首先要将恢复的文件或表空间设为脱机(Offline),但是不包括系统表空间或活动的还原表空间。之后修复(Restore)损坏的操作系统文件,即将备份的物理文件复制回数据库中原来的位置,最后再将写在归档日志文件和联机日志文件的所有提交的数据复原(Recover)。

restore 命令将数据文件带回到过去,即备份的时间点;recover 命令复原从备份到数据文件崩溃这段时间内所有提交的数据改变;最后完成数据库的完全恢复。

recover 命令将所需要的所有提交的数据从归档日志文件和联机重做日志文件写回到需要复原的数据库文件中。recover 命令有三种使用方法:

- recover database。该命令只能在数据库加载(Mount)状态时使用,因为在恢复时 Oracle 要使用控制文件,数据库在 mount 或 open 状态时控制文件才可以访问。
- recover tablespace ＜表空间名＞。该命令只能在数据库打开状态时使用。
- recover datafile ＜数据文件名＞|＜数据文件号＞。该命令在数据库打开或关闭状态时均可以使用。

【实验目的】

（1）了解归档模式下数据库恢复的优缺点。
（2）掌握数据库运行时数据文件破坏的恢复方法。
（3）掌握数据库关闭时数据文件破坏的恢复方法。

【实验内容】

（1）数据库运行时数据文件破坏的数据库恢复。
（2）数据库关闭时数据文件破坏的数据库恢复。

【实验步骤】

1. 数据库运行时数据文件破坏的数据库恢复

数据库在运行时，如果数据文件被破坏，但数据库仍然处于运行状态，可以在数据库运行状态对数据库进行恢复，在整个恢复过程中数据库处于运行状态。数据库运行时，数据文件破坏的数据库恢复步骤如下：

（1）使用数据字典 dba_data_files 获得要恢复的数据文件与对应的表空间及它们的状态信息。

（2）使用数据字典 dba_tablespaces 获得要恢复的表空间的联机状态信息，也可以使用数据字典 v＄datafile 确认要恢复的数据文件是在脱机还是联机状态。

（3）如果要恢复的表空间处在联机状态，要先将该表空间设为脱机状态，使用命令 Alter tablespace ＜表空间名＞ offline。也可以将数据文件设为脱机状态，使用命令 alter database datafile ＜数据文件号＞ offline。

（4）使用操作系统复制命令，将备份的数据文件复制到数据库原来的位置。

（5）使用 recover 命令将所有提交的数据从归档日志文件和重做联机日志文件中重新写入已经修复的数据文件。这里即可以使用命令 recover tablespace ＜表空间名＞，也可以使用命令 Recover datafile ＜数据文件名＞|＜数据文件号＞进行恢复。

（6）当恢复完成后，使用命令 Alter tablespace ＜表空间名＞ online 或命令 alter database datafile ＜数据文件号＞ online 将表空间或数据文件重新设置为联机状态。

下面通过一个联机备份及完全恢复实例演示数据库备份及恢复过程。首先建一个数据表 test，插入一条数据，然后进行数据库备份，这里仅备份 test 表所在的表空间 users。再插入一条数据，该数据显然没有被备份，模拟数据文件损坏，这里通过打开表空间 users 对应的数据文件，插入任意字符，模拟破坏文件。关闭数据库，系统报错，不能正常关闭。将损坏的数据文件脱机，使用操作系统复制命令将备份的数据文件复制到数据库原来的位置。使用 recover 命令将所有提交的数据从归档日志文件和重做联机日志文件中重新写入已经修复的数据文件 users01.dbf。将表空间或数据文件重新设置为联机状态。打开 test 表，查看数据的恢复状态，由于是在归档模式下的恢复，test 表中的两条数据都将被恢复。

建立 test 表,并插入一条数据。

```
SQL> create table test(a int) tablespace users;
表已创建。
SQL> insert into test values(1);
已创建 1 行。
SQL> commit;
提交完成。

SQL> select * from test;
         A
----------
         1
```

使用实验 5.4 数据库备份的知识备份 test 表所在的表空间 users。

设置表空间备份状态。
alter tablespace users begin backup;
备份表空间的数据文件。
host copy E:\Oracle\product\10.2.0\oradata\ORCL\users01.DBF e:\hotback1;
结束表空间备份状态。
alter tablespace users end backup;

继续插入第二条数据 2,此时数据库已完成备份,新增的数据 2 不在备份文件里,但被记录在日志文件中。

```
SQL> insert into test values(2);
已创建 1 行。
SQL> commit;
提交完成。
SQL> select * from test;
         A
----------
         1
         2
```

模拟数据文件 users01.dbf 损坏。在操作系统环境下打开 users01.dbf 文件,任意插入字符,替换原文件,如图 5-13 所示。

保存文件,系统弹出"另存为"对话框,如图 5-14 所示,单击"是"按钮完成替换。

执行 shutdown 命令关闭数据库,系统报错。4 号数据文件 users01.dbf 出错,关闭失败,此时数据库处于打开状态。

```
SQL> shutdown
ORA-01122:数据库文件 4 验证失败
ORA-01110:数据文件 4:'E:\ORACLE\PRODUCT\10.2.0\ORADATA\ORCL\USERS01.DBF'
```

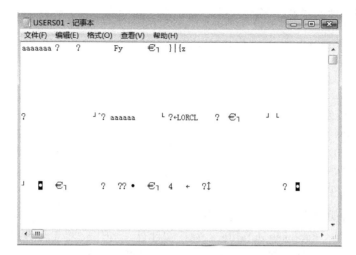

图 5-13　模拟数据文件 users01.dbf 损坏界面

图 5-14　保存文件对话框界面

ORA-01251：读取了文件号 4 的未知文件头部版本

进行恢复。使用数据字典 dba_data_files 获得要恢复的 4 号数据文件与对应的表空间及它们的状态信息。

```
SQL> col file_name for a55
SQL> col tablespace_name for a10
SQL> select file_id,file_name,tablespace_name from dba_data_files;

FILE_ID FILE_NAME                                               TABLESPACE
------- ------------------------------------------------------- ----------
      4 E:\ORACLE\PRODUCT\10.2.0\ORADATA\ORCL\USERS01.DBF       USERS
      3 E:\ORACLE\PRODUCT\10.2.0\ORADATA\ORCL\SYSAUX01.DBF      SYSAUX
      2 E:\ORACLE\PRODUCT\10.2.0\ORADATA\ORCL\UNDOTBS01.DBF     UNDOTBS1
      1 E:\ORACLE\PRODUCT\10.2.0\ORADATA\ORCL\SYSTEM01.DBF      SYSTEM
      5 E:\ORACLE\PRODUCT\10.2.0\ORADATA\ORCL\EXAMPLE01.DBF     EXAMPLE
```

查询数据文件的状态。

```
SQL> select file#,status from v$datafile;

FILE# STATUS
------ -------
     1 SYSTEM
     2 ONLINE
     3 ONLINE
     4 ONLINE
     5 ONLINE
```

查询结果表明,4号数据文件处于联机状态。

使用命令 alter database datafile ＜数据文件号＞ offline 将 4 号数据文件脱机。

```
SQL> alter database datafile 4 offline;
```
数据库已更改。

使用数据字典 v＄datafile 查看数据文件状态信息。

```
SQL> Select file#,status from v$datafile;

FILE# STATUS
------ -------
     1 SYSTEM
     2 ONLINE
     3 ONLINE
     4 RECOVER
     5 ONLINE
```

查询结果表明,4号文件已处于脱机状态。从数据文件备份处复制备份文件到数据库原位置。

```
SQL> Host copy e:\hotback1\users01.dbf
E:\Oracle\product\10.2.0\oradata\orcl\users01.dbf
已复制      1 个文件。
```

此时只恢复了第一条数据1,使用 recover 命令将备份在日志文件中的第二条数据复原。

```
SQL> recover datafile 4;
```
完成介质恢复。

使用命令 alter database datafile ＜数据文件号＞ online 将 4 号数据文件重新设置为联机状态。

```
SQL> alter database datafile 4 online;
```
数据库已更改。

查询数据文件的状态。

```
SQL> select file#,status from v$datafile;

FILE# STATUS
------ -------
    1 SYSTEM
    2 ONLINE
    3 ONLINE
    4 ONLINE
    5 ONLINE
```

查询结果表明,4 号数据文件已处于联机状态。

查看数据表 test 数据,可以看到数据 1、2 都被恢复,其中数据 2 是通过 recover 命令使用日志文件恢复的。

```
SQL> select * from test;
         A
----------
         1
         2
```

2. 数据库关闭时数据文件破坏的数据库恢复

数据库在关闭时,如果数据文件被物理破坏,则数据库不能启动,这时如果启动数据库到 mount 状态自动停止,在这种情况下可以按以下步骤对数据库进行恢复。

(1) 使用 startup mount 命令加载数据库。

(2) 使用数据字典 v$datafile 确认要恢复的数据文件是在脱机还是联机状态。

(3) 如果要恢复的数据文件处在联机状态,要先使用 alter database datafile <数据文件号> offline 命令将该数据文件设为脱机状态。

(4) 使用 alter database open 命令将数据库打开。

(5) 使用操作系统复制命令,将备份的数据文件复制到数据库原来的位置。

(6) 使用 recover 命令将所有提交的数据从归档日志文件和重做联机日志文件中重新写入已经修复的数据文件。这里既可以使用命令 recover tablespace <表空间名>,也可以使用命令 Recover datafile <数据文件名>|<数据文件号>进行恢复。

(7) 当恢复完成后,使用命令 Alter database tablespace <表空间名> online 或命令 alter database datafile <数据文件号> online 将表空间或数据文件重新设置为联机状态。

下面通过一个联机备份及完全恢复实例演示数据库备份及恢复过程。首先建立一个数据表 test,插入一条数据,然后进行数据库备份,这里仅备份 test 表所在的表空间 users。再插入一条数据,该数据显然没有被备份,模拟数据文件损坏,关闭数据库,删除表空间 users 对应的数据文件 users01.dbf。启动数据库时发现系统只能启动到 mount

状态,将损坏的数据文件脱机,使用操作系统复制命令将备份的数据文件 users01.dbf 复制到数据库原来的位置。使用 recover 命令将所有提交的数据从归档日志文件和重做联机日志文件中重新写入已经修复的数据文件。将表空间或数据文件重新设置为联机状态。打开 test 表,查看数据的恢复状态,由于是在归档模式下的恢复,test 表中的两条数据都将被恢复。

如果想控制归档位置,可以使用 alter system set LOG_ARCHIVE_DEST_n 命令设置归档路径。

```
SQL> alter system set log_archive_dest_1="LOCATION=E:\ARCH1 MANDATORY";
```
系统已更改。

建立 test 表,并插入一条数据。

```
SQL> create table test(a int) tablespace users;
```
表已创建。
```
SQL> insert into test values(1);
```
已创建 1 行。
```
SQL> commit;
```
提交完成。

```
SQL> select * from test;
         A
----------
         1
```

使用实验 5.4 数据库备份的知识备份 test 表所在的表空间 users。

设置表空间备份状态。
```
alter tablespace users begin backup;
```
备份表空间的数据文件。
```
host copy E:\Oracle\product\10.2.0\oradata\ORCL\users01.DBF e:\hotback1;
```
结束表空间备份状态。
```
alter tablespace users end backup;
```

继续插入第二条数据 2,此时数据库已完成备份,故新增的数据 2 不在备份文件里,但被记录在日志文件中。

```
SQL> insert into test values(2);
```
已创建 1 行。
```
SQL> commit;
```
提交完成。
```
SQL> select * from test;
         A
----------
         1
         2
```

使用命令 archive log list 查看当前联机日志信息。

```
SQL> archive log list;
数据库日志模式              存档模式
自动存档                    启用
存档终点                    E:\ARCH1
最早的联机日志序列          44
下一个存档日志序列          46
当前日志序列                46
```

查询结果表明,当前日志序列号为 46,即 46 号日志文件存储了 tset 表插入数据 2 的信息。

使用 alter system switch logfile 命令手动切换日志。这里使用该命令进行了三次切换,目的是为了将当前的三组联机日志全部归档,以便在数据库恢复时更清楚地看到 Oracle 是怎样使用归档日志进行恢复的。

```
SQL> alter system switch logfile;
系统已更改。

SQL> alter system switch logfile;
系统已更改。

SQL> alter system switch logfile;
系统已更改。
```

再使用命令 archive log list 查看当前联机日志信息。

```
SQL>  archive log list;
数据库日志模式              存档模式
自动存档                    启用
存档终点                    E:\ARCH1
最早的联机日志序列          47
下一个存档日志序列          49
当前日志序列                49
```

结果表明,从 46 号开始的三个联机日志文件都已得到归档。也可以到操作系统的归档目录 E:\ARCH1 下查看归档日志的信息,如图 5-15 所示。

结果表明,在操作系统的归档目录下,从 46 号开始的 3 个联机日志文件也都得到归档。

模拟数据文件 users01.dbf 损坏。在这里使用删除命令删除数据文件 users01.dbf。当数据库打开时,禁止删除数据文件,所以先关闭数据库。

```
SQL> shutdown immediate;
数据库已经关闭。
已经卸载数据库。
ORACLE 例程已经关闭。
```

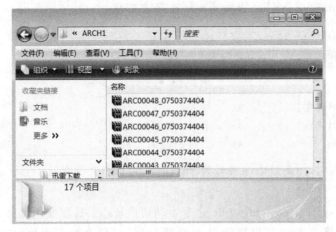

图 5-15   操作系统归档目录信息

```
SQL> host del E:\Oracle\product\10.2.0\oradata\orcl\users01.dbf;
```

此时打开数据库,由于数据文件 users01.dbf 不存在,系统检测到控制文件,只能启动到 mount 状态。

```
SQL> startup
ORACLE 例程已经启动。

Total System Global Area   612368384 bytes
Fixed Size                   1292036 bytes
Variable Size              171968764 bytes
Database Buffers           432013312 bytes
Redo Buffers                 7094272 bytes
数据库装载完毕。
ORA-01157:无法标识/锁定数据文件 4 -请参阅 DBWR 跟踪文件
ORA-01110:数据文件 4:'E:\ORACLE\PRODUCT\10.2.0\ORADATA\ORCL\USERS01.DBF'
```

由于数据库没有打开,要使用数据字典 v$datafile 而不能用数据字典 dba_tablespaces 确认要恢复的数据文件是在脱机还是联机状态。

```
SQL> select file#,status from v$datafile;

     FILE# STATUS
---------- -------
         1 SYSTEM
         2 ONLINE
         3 ONLINE
         4 ONLINE
         5 ONLINE

已选择 5 行。
```

查询结果表明,4号数据文件处于联机状态。

数据库经过关闭状态,动态参数 log_archive_dest_n 已经失效,归档路径 E:\ARCH1 是在线设置的,需要重新使用 alter system set LOG_ARCHIVE_DEST_n 命令设置归档路径。也可将参数 log_archive_dest_n 写在参数文件中,在下一次改变设置前一直有效。

SQL> alter system set log_archive_dest_1="LOCATION=E:\ARCH1 MANDATORY";
系统已更改。

使用命令 alter database datafile ＜数据文件号＞ offline 将 4 号数据文件脱机。

SQL> alter database datafile 4 offline;
数据库已更改。

使用数据字典 v＄datafile 查看数据文件状态信息。

SQL> select file#,status from v$datafile;

```
FILE#   STATUS
------  -------
     1  SYSTEM
     2  ONLINE
     3  ONLINE
     4  OFFLINE
     5  ONLINE
```

已选择 5 行。

查询结果表明,4号文件已处于脱机状态。使用 alter database open 命令将数据库打开。打开数据库后,其他数据文件可以继续使用。

SQL> alter database open;
数据库已更改。

使用操作系统复制命令,将备份的数据文件复制到数据库原来的位置。

SQL> Host copy e:\hotback1\users01.dbf E:\Oracle\product\10.2.0\oradata\orcl\users01.dbf
已复制   1 个文件。

此时只恢复了第一条数据 1,使用 recover 命令将备份在日志文件中的第二条数据复原。

SQL> recover datafile 4;
ORA-00279: 更改 1400292 (在 03/03/2016 21:15:36 生成) 对于线程 1 是必需的
ORA-00289: 建议: E:\ARCH1\ARC00046_0750374404.001
ORA-00280: 更改 1400292 (用于线程 1) 在序列 #46 中

指定日志：{< RET> =suggested | filename | AUTO | CANCEL}

已应用的日志。
完成介质恢复。

上述提示信息表明，要成功恢复该数据文件，只用在线日志中的数据是不够的，必须使用包含该数据文件的归档日志才可以进行数据恢复。此时需要的归档日志是 E:\ARCH1\ ARC00046_0750374404，这是联机日志序列号为 46 号的归档日志，该文件位于系统设定的归档路径 E:\ARCH1 中。

其中 RET 表示按下 Enter 键接受该日志文件，系统提示的目录中存在该日志文件时使用该选项。

filename 表示重新输入归档日志文件名及路径，系统提示的目录中不存在该日志文件时使用该选项。

AUTO 表示自动使用默认路径中的归档日志进行恢复。

CANCEL 表示终止恢复过程，用于不完全数据库恢复，恢复到某一阶段时停止恢复。

如果一个归档日志文件中的数据不足以将数据库文件恢复到当前状态，系统会自动提示输入下一个日志文件名及路径进行恢复，依次进行，直到成功完成恢复，系统会自动停止。

使用命令 alter database datafile ＜数据文件号＞ online 将 4 号数据文件重新设置为联机状态。

```
SQL> alter database datafile 4 online;
```
数据库已更改。

查询数据文件的状态。

```
SQL> select file#,status from v$datafile;

FILE# STATUS
------ -------
     1 SYSTEM
     2 ONLINE
     3 ONLINE
     4 ONLINE
     5 ONLINE
```

查询结果表明，4 号数据文件已处于联机状态。

查看数据表 test 数据。

```
SQL> select * from test;
         A
----------
         1
         2
```

可以看到数据 1、2 都被恢复,其中数据 2 是通过 recover 命令使用日志文件恢复的。

实验 5.6　逻　辑　备　份

　　Oracle Export/Import 工具是一个操作简单、方便灵活的备份恢复和数据迁移工具,它可以实施全库级、用户级、表级的数据备份和恢复。Export 从数据库中导出数据到 dump 文件中。Import 从 dump 文件中导入数据到数据库中。Export 导出的是二进制格式的文件,不可以手工编辑,否则会损坏数据。该文件在 Oracle 支持的任何平台上都是一样的格式,可以在各平台上通用。

Export 和 Import 操作方式:
- 表方式(T):可以将指定的表导入/导出。
- 用户方式(U):可以将指定用户的所有数据对象导入/导出。
- 表空间方式:可以将指定表空间的所有数据对象导入/导出。
- 全库方式(Full):将数据库中的所有对象导入/导出(不包含 SYS 用户的对象)。

【实验目的】

(1) 了解逻辑备份的优缺点。
(2) 掌握逻辑备份的步骤。

【实验内容】

(1) 备份用户指定的表。
(2) 将表从一个用户移到另一个用户。

【实验步骤】

1. 备份用户指定的表

使用 Export/Import 备份 system 的表 dept。

以 system 用户登录数据库,建立数据表 dept,并插入 5 条数据。

```
SQL> connect system/jsj@orcl
已连接。
SQL> create table dept
  2   (deptno number(14) not null,
  3    dname char(20) not null,
  4    loc char(20));

表已创建。
SQL> insert into dept values (10,'销售','大连');
已创建 1 行。
```

```
SQL> insert into dept values (20,'采购','大连');
已创建 1 行。
SQL> insert into dept values (30,'行政','沈阳');
已创建 1 行。
SQL> insert into dept values (40,'人事','大连');
已创建 1 行。
SQL> insert into dept values (50,'售后','沈阳');
已创建 1 行。
SQL> commit;
提交完成。
SQL> select * from dept;
DEPTNO DNAME                LOC
------ -------------------- -------
    10 销售                 大连
    20 采购                 大连
    30 行政                 沈阳
    40 人事                 大连
    50 售后                 沈阳
```

利用 exp 命令备份 dept 表，导出到 e:\back 路径下的 dept.dmp 文件中。

```
SQL> host exp system/jsj@orcl tables=dept file=e:\back\dept.dmp
Export: Release 10.2.0.3.0 -Production on 星期一 3月 7 15:20:01 2016
Copyright (c) 1982, 2005, Oracle. All rights reserved.
连接到: Oracle Database 10g Enterprise Edition Release 10.2.0.3.0 -Production
With the Partitioning, OLAP and Data Mining options
已导出 ZHS16GBK 字符集和 AL16UTF16 NCHAR 字符集
即将导出指定的表通过常规路径...
. . 正在导出表                    DEPT 导出了       5 行
成功终止导出，没有出现警告。
```

结果表明，dept 表的 5 行数据都成功导出了。在操作系统下同样会看到已生成的 dept.dmp 文件，如图 5-16 所示。

图 5-16 操作系统中生成导出 dept.dmp 文件界面

继续往 dept 表中插入第 6 条数据。

```
SQL> insert into dept values (60,'售后','成都');
```
已创建 1 行。
```
SQL> commit;
```
提交完成。
```
SQL> select * from dept;
DEPTNO DNAME                LOC
------ -------------------- -------
    10 销售                 大连
    20 采购                 大连
    30 行政                 沈阳
    40 人事                 大连
    50 售后                 沈阳
    60 售后                 成都
```

已选择 6 行。

现在删除 dept 表,再使用 imp 命令将已备份的 dept 表导入。

```
SQL> drop table dept;
```
表已删除。
```
SQL> host imp system/jsj@orcl tables=dept file=e:\back\dept.dmp
Import: Release 10.2.0.3.0 -Production on 星期一 3月 7 15:40:41 2016
Copyright (c) 1982, 2005, Oracle. All rights reserved.
连接到: Oracle Database 10g Enterprise Edition Release 10.2.0.3.0 -Production
With the Partitioning, OLAP and Data Mining options
经由常规路径由 EXPORT:V10.02.01 创建的导出文件
已经完成 ZHS16GBK 字符集和 AL16UTF16 NCHAR 字符集中的导入
. 正在将 SYSTEM 的对象导入到 SYSTEM
. 正在将 SYSTEM 的对象导入到 SYSTEM
. . 正在导入表              "DEPT"导入了       5 行
成功终止导入,没有出现警告。
```

显示结果表明,只恢复了 5 条数据,即使用 imp 导入时的数据。第 6 条数据是在导出后插入的,不能恢复。也可使用 SQL 命令查询结果进行验证。

```
SQL> select * from dept;
DEPTNO DNAME                LOC
------ -------------------- -------
    10 销售                 大连
    20 采购                 大连
    30 行政                 沈阳
    40 人事                 大连
    50 售后                 沈阳
```

2. 将表从一个用户移到另一个用户

将 system 的 dept 表移到用户 user01。首先新建用户 user01,口令 a,默认表空间

users,使用表空间 users 的额度为 1000KB。

```
SQL> create user user01
  2  identified by a
  3  default tablespace users
  4  temporary tablespace temp
  5  quota 1000k on users;
```
用户已创建。

授予用户 user01 登录数据库及建表的权限。

```
SQL> grant create session,create table to user01;
```
授权成功。

使用 exp 命令将 system 的 dept 表导出。

```
SQL> host exp system/jsj@orcl tables=system.dept file=e:\back\dept.dmp
Export: Release 10.2.0.3.0 -Production on 星期一 3月 7 16:15:23 2016
Copyright (c) 1982, 2005, Oracle. All rights reserved.
连接到: Oracle Database 10g Enterprise Edition Release 10.2.0.3.0 -Production
With the Partitioning, OLAP and Data Mining options
已导出 ZHS16GBK 字符集和 AL16UTF16 NCHAR 字符集
即将导出指定的表通过常规路径...
. . 正在导出表                 DEPT 导出了       5 行
成功终止导出, 没有出现警告。
```

使用数据字典 user_tables 查看 system 的 dept 表所在的表空间。

```
SQL> Select table_name, tablespace_name from user_tables where table_name=
'DEPT';
TABLE_NAME                     TABLESPACE
------------------------------ ----------
DEPT                           SYSTEM
```

结果表明,用户 system 的 dept 表在其默认的系统表空间上。

使用 exp 命令将 system 的 dept 表导入到用户 user01。

```
SQL> host imp system/jsj@orcl fromuser=system touser=user01 tables=dept fi
le=e:\back\dept.dmp

Import: Release 10.2.0.3.0 -Production on 星期一 3月 7 16:23:56 2016
Copyright (c) 1982, 2005, Oracle. All rights reserved.
连接到: Oracle Database 10g Enterprise Edition Release 10.2.0.3.0 -Production
With the Partitioning, OLAP and Data Mining options
经由常规路径由 EXPORT:V10.02.01 创建的导出文件
已经完成 ZHS16GBK 字符集和 AL16UTF16 NCHAR 字符集中的导入
. 正在将 SYSTEM 的对象导入到 USER01
. . 正在导入表              "DEPT" 导入了        5 行
```

成功终止导入，没有出现警告。

结果表明，dept 表已被成功导入。使用 SQL 命令验证 user01 的 dept 表。

```
SQL> select * from user01.dept;
DEPTNO DNAME              LOC
------ ------------------ ------
    10 销售                大连
    20 采购                大连
    30 行政                沈阳
    40 人事                大连
    50 售后                沈阳
```

结果表明，5 条数据已被成功导入到用户 user01。使用数据字典 user_tables 查看 user01 的 dept 表所在的表空间。首先要切换到用户 user01。

```
SQL> conn user01/a@orcl
已连接。
SQL> select table_name, tablespace_name from user_tables where table_name=
'DEPT';
TABLE_NAME                     TABLESPACE
------------------------------ ----------
DEPT                           USERS
```

查询结果表明，dept 表被从原来的表空间 system 导入到 user01 的默认表空间 users。

实验 5.7　数　据　泵

Oracle 10g 引入了数据泵（Data Dump）技术，数据泵导出导入（EXPDP 和 IMPDP）可实现如下功能：实现逻辑备份和逻辑恢复；在数据库用户之间移动对象；在数据库之间移动对象；实现表空间搬移。

【实验目的】

（1）了解数据泵的优缺点。
（2）掌握数据泵的使用方法。

【实验内容】

（1）备份用户指定的表。
（2）不同用户及不同表空间的数据移动。

【实验步骤】

1. 备份用户指定的表

使用 EXPDP 工具时，其转储文件只能被存放在 DIRECTORY 对象对应的 OS 目录中，而不能直接指定转储文件所在的 OS 目录。因此，使用 EXPDP 工具时必须首先建立 DIRECTORY 对象，并且需要为数据库用户授予使用 DIRECTORY 对象权限。使用数据字典 dba_directories 查找当前数据库的 DIRECTORY 对象对应的操作系统目录。

```
set line 120
SQL> col owner for a6
SQL> col directory_name for a20
SQL> col directory_path for a65
SQL> select * from dba_directories;
OWNER DIRECTORY_NAME  DIRECTORY_PATH
----- --------------- -----------------------------------------------------------------
SYS   DPUMP_DIR       e:\back
SYS   SUBDIR          E:\Oracle\product\10.2.0\db_1\demo\schema\order_entry\/2002/Sep
SYS   XMLDIR          E:\Oracle\product\10.2.0\db_1\demo\schema\order_entry\
SYS   MEDIA_DIR       E:\Oracle\product\10.2.0\db_1\demo\schema\product_media\
SYS   LOG_FILE_DIR    E:\Oracle\product\10.2.0\db_1\demo\schema\log\
SYS   WORK_DIR        C:\ADE\aime_vista_ship\Oracle/work
SYS   DATA_FILE_DIR   E:\Oracle\product\10.2.0\db_1\demo\schema\sales_history\
SYS   DATA_PUMP_DIR   E:\Oracle\product\10.2.0\admin\orcl\dpdump\
SYS   ADMIN_DIR       C:\ADE\aime_vista_ship\Oracle/md/admin
```

已选择 9 行。

查询结果表明，当前的 DIRECTORY 对象对应的操作系统目录是 e:\back，如果 DIRECTORY 对象没有设置，系统默认路径是 DATA_PUMP_DIR 对象对应的操作系统目录，从上面查询可以看到，该目录是 E:\Oracle\product\10.2.0\admin\orcl\dpdump\。也可以使用如下命令建立新的 DIRECTORY 对象对应的操作系统目录：

```
SQL> create directory dpump_dir as 'e:\back';
```
目录已创建。

将目录 DPUMP_DIR 的读写权限授予用户 user01。

```
SQL> grant read,write on directory dpump_dir to user01;
```
授权成功。

以 user01 用户登录数据库，建立数据表 dept，并插入 5 条数据。

```
SQL> connect user0/a@orcl
已连接。
SQL> create table dept
  2  (deptno number(14) not null,
  3  dname char(20) not null,
  4  loc char(20));
```

表已创建。

```
SQL> insert into dept values (10,'销售','大连');
已创建 1 行。
SQL> insert into dept values (20,'采购','大连');
已创建 1 行。
SQL> insert into dept values (30,'行政','沈阳');
已创建 1 行。
SQL> insert into dept values (40,'人事','大连');
已创建 1 行。
SQL> insert into dept values (50,'售后','沈阳');
已创建 1 行。
SQL> commit;
提交完成。
SQL> select * from dept;
DEPTNO DNAME              LOC
------ ------------------ ------
    10 销售               大连
    20 采购               大连
    30 行政               沈阳
    40 人事               大连
    50 售后               沈阳
```

利用 expdp 命令备份 dept 表,导出到 e:\back 路径下的 2016dept.dmp 文件中。

```
SQL> Host expdp user01/a@orcl dumpfile=2016dept.dmp directory=dpump_dir
tables=dept
Export: Release 10.2.0.3.0 -Production on 星期二, 08 3 月, 2016 12:19:15
Copyright (c) 2003, 2005, Oracle. All rights reserved.
连接到: Oracle Database 10g Enterprise Edition Release 10.2.0.3.0 -Production
With the Partitioning, OLAP and Data Mining options
启动 "USER01"."SYS_EXPORT_TABLE_01": user01/********@orcl
dumpfile=2016dept.dmp
directory=dpump_dir tables=dept
正在使用 BLOCKS 方法进行估计...
处理对象类型 TABLE_EXPORT/TABLE/TABLE_DATA
使用 BLOCKS 方法的总估计: 64 KB
处理对象类型 TABLE_EXPORT/TABLE/TABLE
. . 导出了 "USER01"."DEPT"                5.804 KB     5 行
已成功加载/卸载了主表 "USER01"."SYS_EXPORT_TABLE_01"
**********************************************************************
USER01.SYS_EXPORT_TABLE_01 的转储文件集为:
  E:\BACK\2016DEPT.DMP
作业 "USER01"."SYS_EXPORT_TABLE_01" 已于 12:19:24 成功完成
```

结果表明,dept 表的 5 行数据都成功导出了。在操作系统下同样会看到已生成的 2016dept.dmp 文件,如图 5-17 所示。

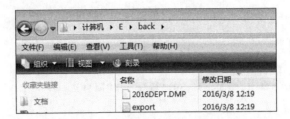

图 5-17　操作系统下生成导出文件界面

继续往 dept 表中插入第 6 条数据。

```
SQL> insert into dept values (60,'售后','成都');
已创建 1 行。
SQL> commit;
提交完成。
SQL> select * from dept;
 DEPTNO DNAME                LOC
------- -------------------- -------
     10 销售                 大连
     20 采购                 大连
     30 行政                 沈阳
     40 人事                 大连
     50 售后                 沈阳
     60 售后                 成都
已选择 6 行。
```

现在删除 dept 表,再使用 impdp 命令将已备份的 dept 表导入。

```
SQL> drop table dept;
表已删除。
SQL> Host impdp user01/a@orcl dumpfile=2016dept.dmp directory=dpump_dir tables=d
ept
Import: Release 10.2.0.3.0 -Production on 星期二, 08 3 月, 2016 12:24:46
Copyright (c) 2003, 2005, Oracle. All rights reserved.
连接到: Oracle Database 10g Enterprise Edition Release 10.2.0.3.0 -Production
With the Partitioning, OLAP and Data Mining options
已成功加载/卸载了主表 "USER01"."SYS_IMPORT_TABLE_01"
启动 "USER01"."SYS_IMPORT_TABLE_01": user01/********@orcl
dumpfile=2016dept.dmp
directory=dpump_dir tables=dept
处理对象类型 TABLE_EXPORT/TABLE/TABLE
处理对象类型 TABLE_EXPORT/TABLE/TABLE_DATA
. . 导入了 "USER01"."DEPT"                           5.804 KB       5 行
作业 "USER01"."SYS_IMPORT_TABLE_01" 已于 12:24:50 成功完成
```

显示结果表明,只恢复了 5 条数据,即使用 impdp 导入时的数据。也可使用 SQL 命令查询结果进行验证。

```
SQL> select * from dept;
DEPTNO  DNAME               LOC
------  ------------------  -------
    10  销售                大连
    20  采购                大连
    30  行政                沈阳
    40  人事                大连
    50  售后                沈阳
```

2. 不同用户及不同表空间的数据移动

将用户 user01 的数据对象移到用户 user02,同时将对象所在的表空间由 users 改为 test。首先切换到 system 用户,新建用户 user02,口令 b,默认表空间 test,使用表空间 test 的额度为 1000KB。

```
SQL> conn system/jsj@orcl
已连接。
SQL> create user user02
  2  Identified by b
  3  default tablespace test
  4  Temporary tablespace temp
  5  quota 1000k on test;
用户已创建。
```

授予用户 user02 登录数据库及建表的权限。

```
SQL> grant create session,create table to user02;
授权成功。
```

将目录 DPUMP_DIR 的读写权限授予用户 user02。

```
SQL> grant read,write on directory dpump_dir to user02;
授权成功。
```

切换到 user01 用户,使用数据字典 user_objects 查看 user01 的全部对象。

```
SQL> col object_name for a35
SQL> Select object_name,object_type,status from user_objects;

OBJECT_NAME                         OBJECT_TYPE         STATUS
----------------------------------  ------------------  -------
BIN$XCIN5zUTRf+ WkbUQNIUAlg==$0     TABLE               VALID
BIN$o1SwuL3JTDKkQgtqsBCz2Q==$0     TABLE               VALID
BIN$0k6lvD8XRSC4qSKpDj71Qw==$0     TABLE               VALID
BIN$SY8n2VtEQCa+ dNxRUBXsUQ==$0    TABLE               VALID
BIN$dUHfYhUTTcakGpt2EHcsOQ==$0     TABLE               VALID
DEPT                                TABLE               VALID
```

已选择 6 行。

显示结果表明，user01 有 6 个数据对象，其中以 BIN 开头的对象是回收站中被删除的 user01 的表，这些表在数据库闪回时被使用。

利用数据字典 user_tables 查看 user01 的表及所在的表空间。

SQL> Select table_name, tablespace_name from user_tables;

```
TABLE_NAME                    TABLESPACE
----------------------------- ----------
DEPT                          USERS
```

显示结果表明，user01 的表都存储在表空间 users 上。

下面使用数据泵进行数据导出的准备工作。在本实验中，由于参数较多，将所有参数存放到一个名为 expdpuser01.txt 的文本文件中，expdp 应用程序调用该文本文件实现一键导出备份。expdpuser01.txt 文件中的内容如下：

```
directory=dpump_dir
dumpfile=user01back.dmp
Schemas=user01
```

将以上内容存放到 E:\back\exp 下的 expdpuser01.txt 文件中，如图 5-18 所示。

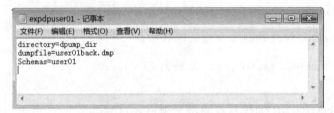

图 5-18 导出批处理文件内容

执行 expdp，调用 expdpuser01.txt 文件，导出数据。

SQL> Host expdp user01/a@orcl parfile=e:\back\exp\expdpuser01.txt
Export: Release 10.2.0.3.0 -Production on 星期三, 09 3月, 2016 11:22:13
Copyright (c) 2003, 2005, Oracle. All rights reserved.
连接到: Oracle Database 10g Enterprise Edition Release 10.2.0.3.0 -Production
With the Partitioning, OLAP and Data Mining options
启动 "USER01"."SYS_EXPORT_SCHEMA_01": user01/********@orcl parfile=e:\back\exp\expdpuser01.txt
正在使用 BLOCKS 方法进行估计...
处理对象类型 SCHEMA_EXPORT/TABLE/TABLE_DATA
使用 BLOCKS 方法的总估计: 64 KB
处理对象类型 SCHEMA_EXPORT/PRE_SCHEMA/PROCACT_SCHEMA
处理对象类型 SCHEMA_EXPORT/TABLE/TABLE

处理对象类型 SCHEMA_EXPORT/TABLE/INDEX/INDEX
处理对象类型 SCHEMA_EXPORT/TABLE/CONSTRAINT/CONSTRAINT
处理对象类型 SCHEMA_EXPORT/TABLE/INDEX/STATISTICS/INDEX_STATISTICS
处理对象类型 SCHEMA_EXPORT/TABLE/COMMENT
. . 导出了 "USER01"."DEPT"   5.804 KB   5 行
已成功加载/卸载了主表 "USER01"."SYS_EXPORT_SCHEMA_01"
****
USER01.SYS_EXPORT_SCHEMA_01 的转储文件集为：
  E:\BACK\USER01BACK.DMP
作业 "USER01"."SYS_EXPORT_SCHEMA_01" 已于 11:23:03 成功完成

显示结果表明，所有的数据导出已经成功，在操作系统目录 e:\back 中可以看到导出的文件 user01back.dmp 已生成，如图 5-19 所示。

图 5-19 操作系统目录下导出文件 **user01back.dmp** 的界面

下面使用数据泵进行数据导入的准备工作。在本实验中，由于参数较多，所有参数存放到一个名为 impdpuser01.txt 的文本文件中，impdp 应用程序调用该文本文件实现一键导入数据。impdpuser01.txt 文件中的内容如下：

```
directory=dpump_dir
dumpfile=user01back.dmp
remap_Schema=user01:user02
remap_tablespace=users:test
```

将以上内容存放到 E:\back\exp 下的 impdpuser01.txt 文件中，如图 5-20 所示。

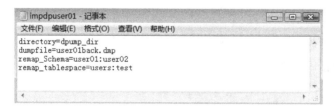

图 5-20 导入批处理文件内容

切换到 user02 用户，执行 impdp，调用 impdpuser01.txt 文件。

SQL> conn user02/b@orcl
已连接。

```
SQL> Host impdp user02/b@orcl parfile=e:\back\exp\impdpuser01.txt

Import: Release 10.2.0.3.0 -Production on 星期四, 10 3月, 2016 10:43:12
Copyright (c) 2003, 2005, Oracle. All rights reserved.
连接到: Oracle Database 10g Enterprise Edition Release 10.2.0.3.0 -Production
With the Partitioning, OLAP and Data Mining options
已成功加载/卸载了主表 "USER02"."SYS_IMPORT_FULL_01"
启动 "USER02"."SYS_IMPORT_FULL_01": user02/********@orcl parfile=e:\back\exp\im
pdpuser01.txt
处理对象类型 SCHEMA_EXPORT/PRE_SCHEMA/PROCACT_SCHEMA
处理对象类型 SCHEMA_EXPORT/TABLE/TABLE
处理对象类型 SCHEMA_EXPORT/TABLE/TABLE_DATA
. . 导入了 "USER02"."DEPT"           5.804 KB       5 行
作业 "USER02"."SYS_IMPORT_FULL_01" 已于 10:43:18 成功完成
```

显示结果表明,user01 的数据对象已被成功地导入到 user02。查看数据字典 user_tables。

```
SQL> Select table_name, tablespace_name from user_tables;

TABLE_NAME                      TABLESPACE
------------------------------  ----------
DEPT                            TEST
```

显示结果表明,在 user02 用户中已导入 dept 表,并且 dept 表所在的表空间为 test。

实验 5.8 数据闪回

使用闪回技术处理更改数据,从根本上改变了恢复技术。恢复错误所花费的时间等于制造错误所花费的时间,它与介质恢复相比,在易用性、可用性和还原时间方面都有明显的优势。

【实验目的】

(1) 了解数据闪回。
(2) 掌握数据闪回使用方法。

【实验内容】

(1) 查询闪回。
(2) 表闪回。
(3) 删除闪回。
(4) 数据库闪回。

【实验步骤】

1. 闪回设置

Oracle 10g 闪回时,系统需工作在归档模式,并开启闪回功能。

确定数据库的归档模式已经开启,使用 SQL 语句在数据字典 v＄database 中查询数据库归档模式。

```
SQL> select name,log_mode from v$database;

NAME        LOG_MODE
---------   ------------
ORCL        NOARCHIVELOG
```

查询结果表明,ORCL 数据库当前工作在非归档模式。更改数据库,使数据库工作在归档模式。

```
SQL> shutdown immediate
数据库已经关闭。
已经卸载数据库。
ORACLE 例程已经关闭。

SQL> startup mount
ORACLE 例程已经启动。

Total System Global Area  612368384 bytes
Fixed Size                  1292036 bytes
Variable Size             255854844 bytes
Database Buffers          348127232 bytes
Redo Buffers                7094272 bytes
数据库装载完毕。

SQL> alter database archivelog;
数据库已更改。

SQL> alter database open;
数据库已更改。

SQL> archive log list;
数据库日志模式          存档模式
自动存档               启用
存档终点               USE_DB_RECOVERY_FILE_DEST
最早的联机日志序列      8
下一个存档日志序列      10
```

当前日志序列   10

查询结果表明,数据库已更改到归档模式。

确认当前闪回模式,使用 SQL 语句在数据字典 v＄database 中查询数据库当前闪回模式。

```
SQL> select flashback_on from v$database;

FLASHBACK_ON
------------------
NO
```

查询结果表明,ORCL 数据库当前关闭闪回模式。更改数据库,使数据库开启闪回模式。

```
SQL> shutdown immediate
SQL> startup mount                    //以 mount 方式启动数据库
SQL> alter database flashback on;     //修改数据库的闪回模式
SQL> alter database open;             //打开数据库

SQL> select flashback_on from v$database;

FLASHBACK_ON
------------------
YES
```

查询结果表明,数据库已开启闪回模式。

2. 查询闪回

在删除表的记录后,希望查询到删除前的值,可以使用查询闪回。创建用户 user02,授予 user02 登录数据库及建表权限,对表空间 users 的使用配额是 200KB。

```
SQL> create user user02 identified by a;
用户已创建。

SQL> grant create session,create table to user02;
授权成功。

SQL> alter user user02 Quota 200k on users;
用户已更改。
```

创建用户 user02 的 test 表,并插入两条数据。

```
SQL> conn user02/a
已连接。
SQL> create table test(a int) tablespace users;
```

表已创建。

```
SQL> insert into test values(3);
```
已创建 1 行。

```
SQL> insert into test values(7);
```
已创建 1 行。

```
SQL> commit;
```
提交完成。

```
SQL> Select * from user02.test;

        A
----------
        3
        7
```

查询系统当前时间，删除 test 表的数据。

```
SQL> select to_char(sysdate,'yyyy-mm-dd hh24:mi:ss') from dual;

TO_CHAR(SYSDATE,'YY
-------------------
2016-04-18 11:16:31

SQL> delete from test;
```
已删除 2 行。

查询 test 表数据。

```
SQL> select * from test;
```
未选定行

查询结果看不到 test 中的数据库。利用闪回查询将数据闪回到时间点 2016-04-18 11:16:31。

```
SQL> select * from test as of timestamp to_timestamp('2016-04-18 11:16:31
  2  ','yyyy-mm-dd hh24:mi:ss');

        A
----------
        3
        7
```

可以看到 test 表中删除前的数据，此时系统只是查询以前的一个快照点而已，并不改变当前表的状态。

3. 表闪回

表闪回利用的是 undo 表空间里记录的被改变前的值,因此,如果表闪回时所需要的 undo 数据由于保留时间超过了初始化参数 undo_retention 所指定的值,从而导致该 undo 数据库数据被其他事务覆盖的话,那么就不能恢复到指定的时间了。

表闪回需要有 flashback any table 的系统权限或者是该表的 flashback 对象权限;需要有该表的 select、insert、delete、alter 权限;必须保证该表启动行移动 row movement。

使用 show 命令查看 undo 相关参数。

```
SQL> show parameter undo

NAME                                 TYPE        VALUE
------------------------------------ ----------- ----------
undo_management                      string      AUTO
undo_retention                       integer     900
undo_tablespace                      string      UNDOTBS1
```

查询结果表明,系统 undo 表空间采用自动管理模式,撤销保持时间为 900s,undo 表空间名为 UNDOTBS1。使用 SQL 命令将撤销保持时间改为 1800s,适当的 undo_retention 设置可避免 undo 数据被覆盖。

```
SQL> alter system set undo_retention=1800;

系统已更改。

SQL> show parameter undo

NAME                                 TYPE        VALUE
------------------------------------ ----------- --------
undo_management                      string      AUTO
undo_retention                       integer     1800
undo_tablespace                      string      UNDOTBS1

SQL> conn user02/a
已连接。
SQL> alter table test enable row movement;
表已更改。
SQL> Select * from test;

    A
----------
    3
    7
```

```
SQL> select to_char(sysdate,'yyyy-mm-dd hh24:mi:ss') from dual;

TO_CHAR(SYSDATE,'YY
-------------------
2016-04-18 11:38:44
```

在当前时间点 2016-04-18 11:38:44，表 test 存在。删除表 test 数据，利用表闪回功能恢复表 test 中的数据到指定时间。

```
SQL> delete from test;
已删除 2 行。
```

此时查看数据，test 表中已没有数据。

```
SQL> Select * from test;
未选定行。
```

利用表闪回功能恢复表 test 中数据。

```
SQL> flashback table test to timestamp to_timestamp('2016-04-18 11:38:44
  2  ','yyyy-mm-dd hh24:mi:ss');

闪回完成。

SQL> Select * from test;

         A
----------
         3
         7
```

也可以将表闪回到指定的 scn，查看在当前系统的 scn，在这个 scn 点表 test 存在。删除表 test 数据。检查数据，确认数据删除，利用表闪回功能恢复表 test 中的数据到指定的 scn。

```
SQL> conn sys/jsj as sysdba
已连接。
SQL> select current_scn from v$database;

CURRENT_SCN
-----------
     738909
SQL> select * from user02.test;
         A
----------
         3
         7
```

```
SQL> conn user02/a@orcl
SQL> delete from test;
已删除 2 行。

SQL> Select * from test;
未选定行。

SQL> flashback table test to scn 738909;
闪回完成。

SQL> select * from test;
         A
----------
         3
         7
```

4. 删除闪回

删除闪回为删除 oracle 10g 中的数据库实体提供了一个安全机制,当用户删除一个表时,oracle 10g 会将该表放到回收站中,回收站中的对象一直会保留,直到用户决定永久删除它们或出现 undo 表空间不足时才会被删除。回收站是一个虚拟容器,用于存储所有被删除的对象。回收站中的数据存放在 undo 表空间上。

```
SQL> conn user02/a
已连接。
SQL> Select * from test;

         A
----------
         3
         7

SQL> drop table test;
表已删除。
```

使用 show 命令显示回收站信息。

```
SQL> show recyclebin;
ORIGINAL NAME    RECYCLEBIN NAME                  OBJECT TYPE  DROP TIME
---------------- -------------------------------- ------------ -------------------
TEST             BIN$rTV/Bo/mSyymiH8zbYRTjw==$0   TABLE        2016-04-18:12:08:14
```

回收站中以 BIN 开头的表就是被删除的表。使用 SQL 语句测试被删除的 test 表。

```
SQL> Select * from test;
Select * from test
```

```
            *
第 1 行出现错误：
ORA-00942：表或视图不存在
```

显示结果表明，test 表已经不存在了，如果此时发现删除这个表是错误的，可以使用删除闪回恢复。

```
SQL> flashback table test to before drop;
```

闪回完成。

```
SQL> Select * from test;

         A
----------
         3
         7
```

显示结果表明，使用删除闪回，test 表被成功恢复了。

真正删除某一个表，而不进入回收站，可以在删除表时增加 purge 选项，此时已不能使用删除闪回恢复数据表了。如将 test 表彻底删除，使用 purge 选项，显示回收站信息，将没有数据行。也可以使用 purge recyclebin 命令清空回收站。

```
SQL> drop table test purge;
表已删除。

SQL> show recyclebin;
SQL>
SQL> purge recyclebin;
```

回收站已清空。

5．数据库闪回

数据库闪回是当数据库出现逻辑错误时，能够将整个数据库回退到出错前的某个时间点上，闪回数据库的日志文件不是由传统的 Log Writer（LGWR）进程写入，而是由 RecoVery WRiter（RVWR）的进程写入，闪回日志文件由 RCWR 进程在恢复区中自动创建和维护。

实现闪回数据库的基础是闪回日志，配置了闪回数据库，系统就会自动创建闪回日志。此时如果数据库里的数据发生变化，oracle 就会将数据被修改前的旧值保存在闪回日志里，当需要闪回数据库时，oracle 就会读取闪回日志里的记录，并应用到数据库上，从而将数据库回退到历史的某个时间点。

闪回恢复区主要通过三个初始化参数来设置和管理。
- db_recovery_file_dest：指定闪回恢复区的位置。

- db_recovery_file_dest_size：指定闪回恢复区的可用空间大小。
- db_flashback_retention_target：指定数据库可以回退的时间,单位为分钟,默认为 1440 分钟。当然,实际上可回退的时间还决定于闪回恢复区的大小,因为里面保存了回退所需要的 flash log。所以这个参数要和 db_recovery_file_dest_size 配合设置。

使用 show 命令显示 db_recovery_file_dest 信息。

```
SQL> show parameter db_recovery_file_dest
NAME                     TYPE         VALUE
------------------------ ------------ ------------------------------
db_recovery_file_dest    string       E:\oracle\product\10.2.0\flash_recovery_area
db_recovery_file_dest_size big integer  2G
```

可以使用 SQL 命令修改 db_recovery_file_dest 和 db_recovery_file_dest_size 参数,如将 db_recovery_file_dest_size 修改成 3G。

```
SQL> alter system set db_recovery_file_dest_size=3g;
系统已更改。

SQL> show parameter db_recovery_file_dest
NAME                     TYPE         VALUE
------------------------ ------------ ------------------------------
db_recovery_file_dest    string       E:\oracle\product\10.2.0\flash_recovery_area
db_recovery_file_dest_size big integer  3G
```

查询结果表明,db_recovery_file_dest_size 参数已修改成 3G。

使用 show 命令显示 db_flashback_retention_target 信息。

```
SQL> show parameter db_flashback_retention_target;

NAME                             TYPE        VALUE
-------------------------------- ----------- ----------
db_flashback_retention_target    integer     1440
```

显示结果表明,当前回退时间是 1400 分钟,可以使用 SQL 命令修改 db_flashback_retention_target 的值,如改成 2880 分钟。

```
SQL> alter system set db_flashback_retention_target=2880;
系统已更改。
```

闪回数据库使用 flashback database 命令,如闪回数据库到指定的 scn。

```
SQL> flashback database to scn 172479
```

6. Flash Version Query 与 Flashback Transaction Query 的应用

Oracle Flashback Version Query 利用保存的回滚信息,可以看到指定的表在某时间段内的任何修改,如电影的回放一样,可以了解表在该期间的任何变化。

Oracle Flashback Transaction Query 确保检查数据库的任何改变在一个事务级别,可以利用此功能进行诊断问题、性能分析和恢复事务。

Flashback version query 依赖 AUM(Auto Undo Management),提供了一个查看行改变的功能,能找到所有已经提交了的行的记录,分析出过去时间都执行了什么操作。Flashback version query 采用 VERSIONS BETWEEN 子句进行查询,常用的方法包括基于系统改变号 VERSIONS_SCN 和基于时间戳 VERSIONS_TIMESTAMP 的查询。

用户 user02 建立 emp 表,插入 emp 表的三条记录。

```
SQL> conn user02/a
已连接。
SQL> create table emp
  2  (empno char(10) not null,
  3   ename char(20) not null,
  4   sal smallint,
  5   comm smallint,
  6   job char(20),
  7   hiredate date,
  8   deptno number(14)
  9  );

表已创建。

SQL> INSERT INTO emp VALUES ('0001','张蓓',1800,500,'经理',sysdate,'10');
已创建 1 行。

SQL> INSERT INTO emp VALUES ('0003','黄欣懿',2800,500,'职员',sysdate,'20');
已创建 1 行。

SQL> INSERT INTO emp VALUES ('0004','邓瑞峰',3800,500,'职员',sysdate,'30');
已创建 1 行。

SQL> commit;

提交完成。

SQL> select * from emp;
EMPNO      ENAME           SAL     COMM  JOB     HIREDATE       DEPTNO
------------------------------------------------------------------------
0001       张蓓            1800    500   经理    19-4月-16       10
0003       黄欣懿          2800    500   职员    19-4月-16       20
0004       邓瑞峰          3800    500   职员    19-4月-16       30
```

事务 1 将 0001 号员工的工资改成 5000,事务 2 将 0003 号员工的工资改成 7000,事务 3 将 0004 号员工的工资改成 9000。现发现事务 2 的修改错误，需要撤销。如果用表闪回恢复到事务 2 之前，事务 3 的修改也撤销了。这时利用 Flash Version Query 与 Flashback Transaction Query 可实现只撤销事务 2 的操作。首先利用 Flashback Version Query 把事务 1、事务 2、事务 3 的操作记录下来，并详细地查询出对表进行的具体操作。

```
SQL> update emp set sal=5000 where empno=0001;
已更新 1 行。
SQL> commit;
提交完成。


SQL> update emp set sal=7000 where empno=0003;
已更新 1 行。
SQL> commit;
提交完成。


SQL> update emp set sal=9000 where empno=0004;
已更新 1 行。
SQL> commit;
提交完成。
```

利用 Flashback Version Query 查询三条修改操作对应的事务标识符 xid。

```
SQL> select versions_xid,versions_operation,sal from emp versions
  2  between timestamp minvalue and maxvalue
  3  order by versions_starttime;


VERSIONS_XID      V  SAL
----------------  -  -----------------------------
070003005F010000  I  2800
070003005F010000  I  3800
070003005F010000  I  1800
060026006E010000  U  5000
0800160072010000  U  7000
06002A006E010000  U  9000


已选择 6 行。
```

查询结果表明，事务 2 的更新操作对应的事务标识符是 0800160072010000，在此时间点，事务执行的 versions_operation 是 updata，将 sal 改为 7000。在上述查询中，versions_starttime、versions_endtime、versions_xid、versions_operation 是伪列。还有一些伪列，如 versions_startscn 和 versions_endscn 显示了该时刻的系统更改号，列 versions_xid 显示了更改该行的事务标识符。

除了使用 timestamp minvalue and maxvalue 分析的修改外，还可以根据需要查询指定时间段内的修改，如显示在 2015-05-07，时间在 15：30 到 16：30 之间 empt 表的所有变更。

```
SQL> select id from test versions between timestamp to_date('2015-05-07 15:30:
00','yyyy-mm-dd hh24:mi:ss') and to_date('2015-05-07 16:30:00','yyyy-mm-dd
hh24:mi:ss')
```

Flashback Version Query 可以审计一段时间内表的所有改变，但是只能发现问题，对于错误的事务不能够处理。Oracle Flashback Transaction Query 是 Flashback Version Query 查询的一个扩充，Flashback Transaction Query 提供了从 FLASHBACK_TRANSACTION_QUERY 视图中获得事务的历史及 Undo_sql（回滚事务对应的 sql 语句），可以审计一个事务的操作并回滚已提交的事务。

如把事务 2 的操作撤销，可使用 Flash Version Quer 得到事务标识符 xid，把该标识符赋予 Flashback Version Query 查询，得到事务 2 的历史操作及回滚的 Undo_sql 操作。

```
SQL> conn sys/jsj as sysdba
已连接。
SQL> select operation,undo_sql from FLASHBACK_TRANSACTION_QUERY where xid=
'08001600720l0000';

OPERATION   UNDO_SQL
-----------------------------------------------------------------------------
UPDATE update "USER02"."EMP" set "SAL"='2800' where ROWID='AAAM84AAEAAAAGkAAB';
BEGIN
```

从显示结果可以看到，在事务 2 中事务执行的是 UPDATE 操作，sal 的原始值是 2800。如果想撤销事务 2 的操作，执行这条 UNDO_SQL。

```
SQL> update "USER02"."EMP" set "SAL"='2800' where ROWID='AAAM84AAEAAAAGkAAB';
已更新 1 行。
SQL> select * from user02.emp;
EMPNO       ENAME           SAL         COMM  JOB    HIREDATE      DEPTNO
-----------------------------------------------------------------------------
0001        张蓓            5000        500   经理   19-4月 -16     10
0003        黄欣懿          2800        500   职员   19-4月 -16     20
0004        邓瑞峰          9000        500   职员   19-4月 -16     30
```

显示结果表明，事务 2 的事务操作已成功撤销。

第 6 章 手工建立数据库

本实验内容是 Oracle 数据库实践训练的一个综合练习。手工建立数据库主要包括以下几个步骤：
(1) 建立数据库需要的相关文件夹。
(2) 准备参数文件。
(3) 创建实例。
(4) 创建数据库。
(5) 配置监听及网络服务。

实验 6.1 手工建立数据库

【实验目的】

(1) 了解 Oracle 的物理结构、逻辑结构及实例之间的关系。
(2) 掌握手工建立数据库的方法。

【实验内容】

(1) 创建数据库实例。
(2) 手工建立数据库。
(3) 配置数据库监听和网络服务。

【实验步骤】

1. 建立数据库 dltest 需要的相关文件夹

注：数据库名字不能超过 8 个字符，不能以数字开头。

```
E:\Oracle\product\10.2.0\oradata\dltest
E:\Oracle\product\10.2.0\admin\dltest\bdump
E:\Oracle\product\10.2.0\admin\dltest\cdump
```

```
E:\Oracle\product\10.2.0\admin\dltest\udump
E:\Oracle\product\10.2.0\admin\dltest\pfile
```

2．准备参数文件

需要创建两个参数文件：init.ora 和 init＋实例名.ora。参数文件的路径如下：

```
E:\Oracle\product\10.2.0\admin\ dltest \pfile\init.ora
E:\Oracle\product\10.2.0\Db_1\database\ init+实例名.ora
```

init.ora 和 init＋实例名.ora 的内容是相同的，数据库启动先检测 init＋实例名.ora，如果没有再检测 init.ora 文件。init＋实例名.ora 中的语句是：

```
ifile='E:\Oracle\product\10.2.0\admin\ dltest \pfile\init.ora'
```

参数文件 init.ora 存放数据库两百多个参数，这里做简化处理，配置部分静态参数和需要配置的动态参数。

```
db_name=dltest                                          //数据库名
db_domain=jsjzy.cn                                      //数据库域名
control_files=('E:\Oracle\product\10.2.0\oradata\dltest\db01.ctl',
'E:\Oracle\product\10.2.0\oradata\dltest\db02.ctl')     //两个镜像控制文件位置
db_cache_size=12M                                       //数据缓冲区
shared_pool_size=64M                                    //共享池
log_buffer=1048576                                      //日志缓冲区
large_pool_size=0                                       //大缓存池
java_pool_size=0                                        //Java 池
workarea_size_policy=auto                               //pga 启用自动内存管理
pga_aggregate_target=12M                                //PGA
undo_management=auto                                    //自动撤销管理
```

在操作系统下用记事本打开参数文件 E:\Oracle\product\10.2.0\admin\ dltest \pfile\init.ora ，如图 6-1 所示。

图 6-1　参数文件内容

3. 创建实例

打开 DOS 命令窗口,以下命令在 DOS 下运行。使用 Oradim 命令创建实例 dltest。

```
C:\Users\user>set Oracle_sid=dltest
C:\Users\user>oradim -new -sid dltest -intpwd jsj -startmode manual
实例已创建。
```

若删除实例,语句为 oradim -delete -sid 实例名。

Oradim 命令创建了 dltest 实例,SYS 口令是 jsj,启动方式是手动。此时右击"我的电脑"图标,在弹出的快捷菜单中选择"管理"命令,在打开的"计算机管理"窗口中选择"服务和应用程序"节点,然后选择"服务"选项打开操作系统服务,看到 dltestserver 服务已生成,如图 6-2 所示。

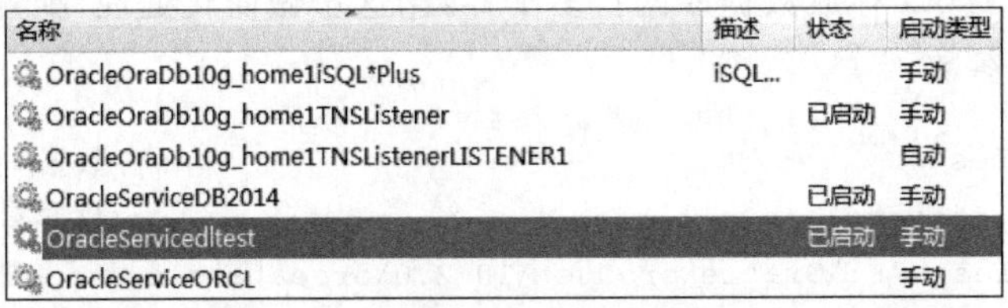

图 6-2 启动 dltestserver 服务界面

4. 创建数据库

```
C:\>sqlplus/nolog
SQL>conn sys/jsj@orcl as sysdba
```

根据参数文件启动数据库实例 dltest。

```
SQL> startup nomount pfile=E:\Oracle\product\10.2.0\admin\dltest\pfile\init.ora
ORACLE 例程已经启动。

Total System Global Area   83886080 bytes
Fixed Size                  1289028 bytes
Variable Size              67110076 bytes
Database Buffers           12582912 bytes
Redo Buffers                2904064 bytes
```

使用 create database 命令创建数据库。

```
SQL> create database dltest
  2  datafile 'e:\Oracle\product\10.2.0\oradata\dltest\system01.dbf' size
     200m reuse autoextend on next 10240k maxsize unlimited
  3  sysaux datafile 'e:\Oracle\product\10.2.0\oradata\dltest\sysaux01.dbf' size
     50m autoextend on next 10m maxsize unlimited
```

```
    4  logfile
    5  group 1 ('e:\Oracle\product\10.2.0\oradata\dltest\undo11.log')size 4m,
    6  group 2 ('e:\Oracle\product\10.2.0\oradata\dltest\undo21.log')size 4m
    7  undo tablespace undo1 datafile
       'e:\Oracle\product\10.2.0\oradata\dltest\undo01.dbf'size 2m reuse
       autoextend on
    8  character set ZHS16GBK;
```
数据库已创建。

数据库初始建立了 system01.dbf 文件对应的系统表空间；sysaux01.dbf 文件对应的系统扩展表空间；undo11.log 文件和 undo21.log 文件对应的两个日志文件组，每组一个成员；undo01.dbf 文件对应的回滚表空间。

在操作系统下打开 dltest 路径，可以看到生成的物理文件，如图 6-3 所示。

图 6-3　dltest 数据库的物理文件

结果表明，生成了两个镜像的控制文件 DB01.CTL 和 DB02.CTL，三个数据文件 SYSAUX01.DBF、SYSTEM01.DBF 和 UNDO01.DBF，两个联机日志文件 UNDO11.LOG 和 UNDO21.LOG。

加载常用的数据字典包，建立数据字典。

```
SQL>conn sys/jsj@orcl as sysdba
SQL>@E:\Oracle\product\10.2.0\db_1\RDBMS\ADMIN\catalog.sql;
...
```

加载 PL/SQL 程序包。

```
SQL>@E:\Oracle\product\10.2.0\db_1\RDBMS\ADMIN\catproc.sql;
...
```

加载环境文件程序包，必须以 system 用户登录运行环境文件程序包。

```
SQL>conn system/manager@orcl
@E:\Oracle\product\10.2.0\db_1\sqlplus\admin\pupbld.sql;
...
```

5. 配置监听及网络服务

数据库创建以后需进行网络配置,系统才可以和新建的数据库连接。网络配置包括两部分:监听配置文件 LISTENER. ORA 及网络连接解析配置文件 TNSNAMES. ORA 的设置。

监听配置文件 LISTENER. ORA 的目录为 E:\Oracle\product\10.2.0\db_1\network\ADMIN\LISTENER. ORA,用记事本打开。可以看到,监听的数据库均在这个文件里,可复制一个已经设置好的数据库程序块,将全局数据库名 GLOBAL_DBNAME 改为 dltest. jsjzy. cn,实例名 SID_NAME 改为 dltest,如图 6-4 所示。

图 6-4 监听文件 LISTENER. ORA 设置界面

网络连接解析配置文件 TNSNAMES. ORA 的目录为 E:\Oracle\product\10.2.0\db_1\network\ADMIN\TNSNAMES. ORA,可用记事本打开。系统中所有数据库的网络连接配置均在这个文件里,可复制一个已经设置好的数据库程序块,服务名 SERVICE_NAME 改为 dltest. jsjzy. cn,如图 6-5 所示。

监听及网络服务配置成功后,在操作系统服务里重启监听程序。右击"我的电脑"图标,从弹出的快捷菜单中选择"管理"命令,在打开的"计算机管理"窗口中选择"服务和应用程序"节点,然后选择"服务"选项,右击 OracleOraDb10g_home1TNSListener,从弹出的快捷菜单中选择"重新启动"命令,如图 6-6 所示。

这时可以启动数据库 dltest 了。@连接符后输入全局数据库名 dltest. jsjzy. cn。

```
C:\Users\user>sqlplus/nolog
SQL*Plus: Release 10.2.0.3.0 -Production on 星期五 4月 1 17:09:27 2016
Copyright (c) 1982, 2006, Oracle. All Rights Reserved.
```

```
tnsnames - 记事本
文件(F) 编辑(E) 格式(O) 查看(V) 帮助(H)
# tnsnames.ora Network Configuration File: E:\oracle\product\10.2.0\db_
# Generated by Oracle configuration tools.

DLTEST.JSJZY.CN =
  (DESCRIPTION =
    (ADDRESS_LIST =
      (ADDRESS = (PROTOCOL = TCP)(HOST = localhost)(PORT = 1521))
    )
    (CONNECT_DATA =
      (SERVICE_NAME = dltest.jsjzy.cn)
    )
  )

ORCL =
  (DESCRIPTION =
    (ADDRESS_LIST =
      (ADDRESS = (PROTOCOL = TCP)(HOST = localhost)(PORT = 1521))
    )
    (CONNECT_DATA =
      (SERVER = DEDICATED)
      (SERVICE_NAME = orcl)
    )
  )

EXTPROC_CONNECTION_DATA =
  (DESCRIPTION =
    (ADDRESS_LIST =
      (ADDRESS = (PROTOCOL = IPC)(KEY = EXTPROC1))
    )
    (CONNECT_DATA =
      (SID = PLSExtProc)
      (PRESENTATION = RO)
    )
  )
```

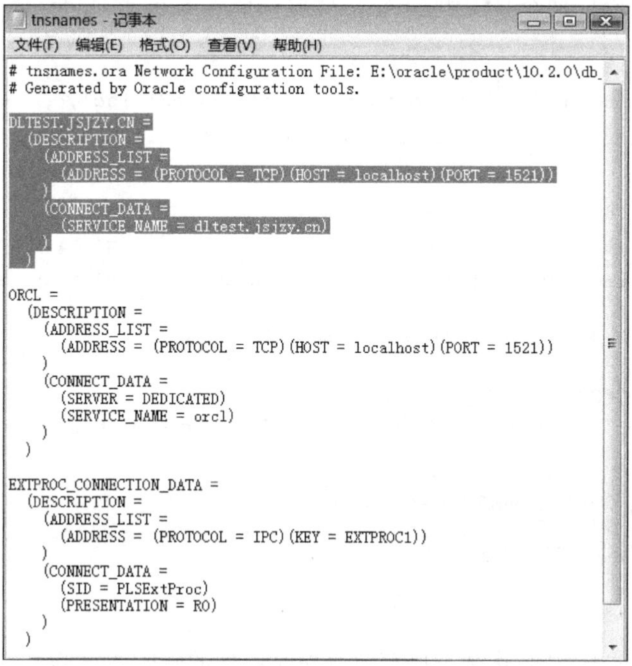

图 6-5 网络连接文件 TNSNAMES. ORA 配置界面

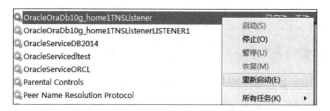

图 6-6 监听程序启动界面

```
SQL> conn sys/jsj@dltest.jsjzy.cn as sysdba
已连接到空闲例程。
SQL> startup
ORACLE 例程已经启动。

Total System Global Area 83886080 bytes
Fixed Size                1289028 bytes
Variable Size            67110076 bytes
Database Buffers         12582912 bytes
Redo Buffers              2904064 bytes
数据库装载完毕。
数据库已经打开。
```

可以看到，数据库 dltest 成功打开。初始建库时只有两个日志组，每组一个成员，为增加安全性，为数据库 dltest 增加一个日志组 group 3，该组包含两个成员 undo31.log 和

undo32.log，为已有的两个日志组，每组再增加一个成员 undo12.log 和 undo22.log。

```
alter database add logfile group 3
('E:\Oracle\product\10.2.0\oradata\dltest\undo31.log',
'E:\Oracle\product\10.2.0\oradata\dltest\undo32.log') size 4m;

alter database add logfile member
'E:\Oracle\product\10.2.0\oradata\dltest\undo12.log' to group 1,
'E:\Oracle\product\10.2.0\oradata\dltest\undo22.log' to group 2;
```

建立一个用户表空间 user001，分流存储数据。

```
create tablespace user001
datafile 'E:\Oracle\product\10.2.0\oradata\dltest\user001.dbf' size 10m reuse
autoextend on next 10m maxsize 100m;
```

设置默认表空间为用户表空间 user001。

```
alter database default tablespace user001;
```

需要建立临时表空间，供数据库存放排序时产生的临时数据等。

```
create TEMPORARY TABLESPACE TEMP TEMPFILE
'E:\Oracle\product\10.2.0\oradata\dltest\temp01.dbf' size 50m
autoextend on next 50m maxsize unlimited;
```

设置系统默认临时表空间。

```
alter database default TEMPORARY tablespace TEMP;
```

检查当前物理结构，查看数据文件。

```
SQL> col tablespace_name for a15
SQL> col file_name for a55
SQL> select tablespace_name,autoextensible,file_name from dba_data_files;

TABLESPACE_NAME AUT FILE_NAME
--------------- --- -------------------------------------------------------
SYSTEM          YES E:\ORACLE\PRODUCT\10.2.0\ORADATA\DLTEST\SYSTEM01.DBF
UNDO1           YES E:\ORACLE\PRODUCT\10.2.0\ORADATA\DLTEST\UNDO01.DBF
SYSAUX          YES E:\ORACLE\PRODUCT\10.2.0\ORADATA\DLTEST\SYSAUX01.DBF
USER001         YES E:\ORACLE\PRODUCT\10.2.0\ORADATA\DLTEST\USER001.DBF
```

查询结果表明，数据库包含 4 个数据文件，与前面操作的结果一致。

查询控制文件。

```
SQL> select name from v$controlfile;
NAME
--------------------------------------------------
```

```
E:\ORACLE\PRODUCT\10.2.0\ORADATA\DLTEST\DB01.CTL
E:\ORACLE\PRODUCT\10.2.0\ORADATA\DLTEST\DB02.CTL
```

查询结果表明,数据库 dltest 包含两个控制文件,与参数文件设置的完全一致。
查询日志文件。

```
SQL> col TYPE for a10
SQL> col MEMBER for a60
SQL> SELECT GROUP#,TYPE,MEMBER FROM v$logfile;
GROUP#   TYPE     MEMBER
------------------------------------------------------------
1        ONLINE   E:\ORACLE\PRODUCT\10.2.0\ORADATA\DLTEST\UNDO11.LOG
2        ONLINE   E:\ORACLE\PRODUCT\10.2.0\ORADATA\DLTEST\UNDO21.LOG
3        ONLINE   E:\ORACLE\PRODUCT\10.2.0\ORADATA\DLTEST\UNDO31.LOG
3        ONLINE   E:\ORACLE\PRODUCT\10.2.0\ORADATA\DLTEST\UNDO32.LOG
1        ONLINE   E:\ORACLE\PRODUCT\10.2.0\ORADATA\DLTEST\UNDO12.LOG
2        ONLINE   E:\ORACLE\PRODUCT\10.2.0\ORADATA\DLTEST\UNDO22.LOG
已选择 6 行。
```

查询结果表明,日志文件分为三组,每组两个成员,与设置的完全一致。
在操作系统路径下也可以看到数据库相应的文件,如图 6-7 所示。

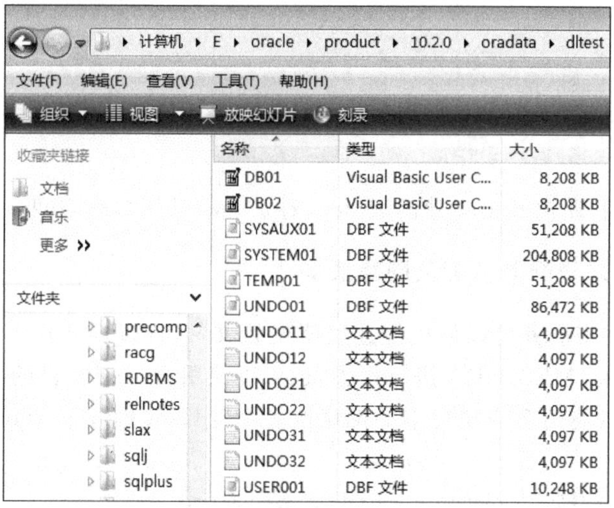

图 6-7 数据库 dltest 对应的物理文件

检验数据库 dltest,新建一个用户 dluser,授予用户 dluser 登录数据库及建表的权限,并有 1000KB 使用表空间 user001 的配额。

```
SQL> create user dluser identified by a;
用户已创建。
SQL> grant create session,create table to dluser;
授权成功。
SQL> alter user dluser
```

```
  2  quota 1000k on user001;
```
用户已更改。

切换到用户 dluser,建立表 dept。

```
SQL> conn dluser/a@dltest.jsjzy.cn
已连接。
SQL> create table dept
  2  (deptno number(14) not null,
  3  dname char(20) not null,
  4  loc char(20),
  5  primary key (deptno));
```

表已创建。

检测表 dept 所在的表空间等信息。

```
SQL> col OWNER for a10
SQL> col TABLESPACE_NAME for a10
SQL> col TABLE_NAME for a10
SQL> select OWNER,TABLESPACE_NAME,TABLE_NAME from all_tables where TABLE_NAME='D
EPT';

OWNER      TABLESPACE TABLE_NAME
---------- ---------- ----------
DLUSER     USER001    DEPT
```

查询结果表明,用户 dluser 建的表 dept 存放在默认用户表空间 user001 上。

6. 使用配置向导配置监听及网络服务

配置监听及网络服务可以使用上述监听配置文件 LISTENER.ORA 及网络连接解析配置文件 TNSNAMES.ORA 进行设置,也可以使用 Oracle 提供的图形配置工具。

网络服务可以使用客户端自带的网络配置向导(Net Configuration Assistant)进行配置。

在"开始"菜单中启动 Net Configuration Assistant,如图 6-8 所示。选择"本地 Net 服务名配置"单选按钮,如图 6-9 所示。

单击"下一步"按钮,可以对本地网络服务名进行添加、删除、测试是否正常连接等操作,选择"添加"单选按钮,如图 6-10 所示。

单击"下一步"按钮,填写服务名,该服务名为连接到 Oracle 数据库的全局数据库名,本例是 dltest.jsjzy.cn,如图 6-11 所示。

图 6-8 Net Configuration Assistant
       选择界面

第 6 章 手工建立数据库

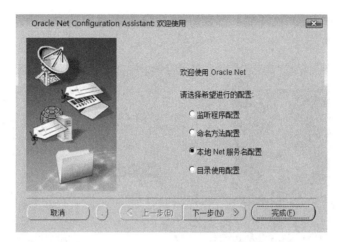

图 6-9　Net Configuration Assistant 启动界面

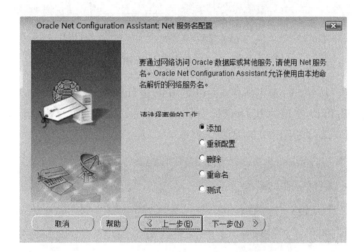

图 6-10　Net Configuration Assistant 添加服务界面

图 6-11　配置服务名界面

单击"下一步"按钮,选择服务需要的协议,默认是 TCP 协议。推荐使用默认的 TCP 协议,如图 6-12 所示。

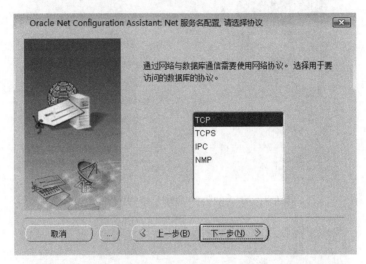

图 6-12　配置协议界面

单击"下一步"按钮,输入主机名,主机名可以是计算机名称,也可以是一个 IP 地址。主机如果是本机,可以使用本机计算机名称 localhost、127.0.0.1,或者本机的 IP 地址。端口号取默认值 1521,如图 6-13 所示。

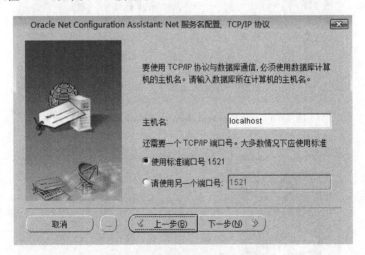

图 6-13　主机名和端口配置界面

单击"下一步"按钮,选择"是,进行测试"单选按钮,进入测试界面,如图 6-14 所示。

在测试时,默认采用的用户名和密码是 system/manager。如果用户 system 的密码不是 manager,测试通不过,选择更改登录后输入正确的用户名和密码,如图 6-15 所示,再进行测试即可,如图 6-16 所示。

测试成功后,单击"下一步"按钮,如图 6-17 所示。这一步是为本地网络服务命名,可以任意取名,本例为全局数据库名 dltest.jsjzy.cn。

第 6 章 手工建立数据库 117

图 6-14 测试选择界面

图 6-15 更改登录界面

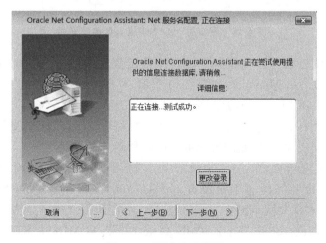

图 6-16 测试成功界面

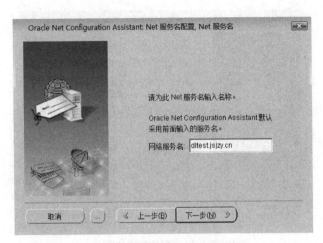

图 6-17　网络服务命名配置界面

单击"下一步"按钮,进入图 6-18 所示界面,询问是否配置另一个 Net 服务名,选择"否"单选按钮,然后单击"下一步"按钮,配置就完成了,如图 6-19 所示。打开 E:\Oracle\product\10.2.0\db_1\network\ADMIN\TNSNAMES.ORA 文件,可以看到和图 6-5 的结果一样。

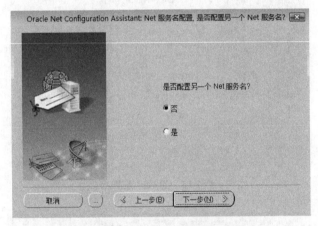

图 6-18　配置另一个 Net 服务名选择界面

监听数据库配置也可以使用客户端自带的配置向导(Net Manager)进行配置。在"开始"菜单中启动 Net Manager,如图 6-20 所示。

进入 Oracle Net Manage 欢迎界面,如图 6-21 所示。

进入 Oracle Net Manage 配置界面,选择监听程序 LISTENER,如图 6-22 所示。在"监听位置"下拉列表中选择"数据库服务"选项,如图 6-23 所示。

进入数据库配置界面,单击"添加数据库"按钮,出现数据库配置对话框,如图 6-24 所示。在"全局数据库名"文本框中输入 dltest.jsjzy.cn,在 SID 文本框中输入 dltest。

选择"文件"→"保存网络配置"命令,完成数据库 dltest 监听的配置,如图 6-25 所示。此时打开 E:\Oracle\product\10.2.0\db_1\network\ADMIN\LISTENER.ORA,可以看到和图 6-5 一样的结果。

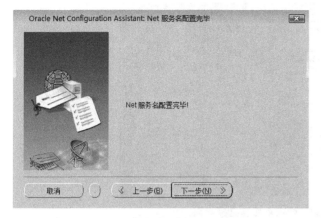

图 6-19 服务名配置完毕界面

图 6-20 在"开始"菜单启动 Net Manage

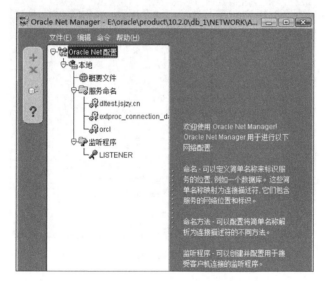

图 6-21 Oracle Net Manage 欢迎界面

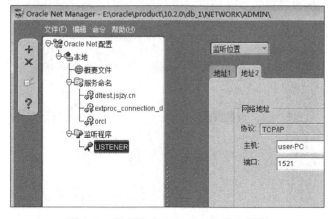

图 6-22 监听程序 LISTENER 选择界面

图 6-23 数据库服务选择界面

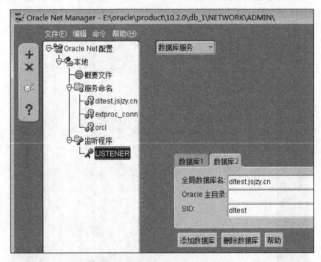

图 6-24 数据库配置界面

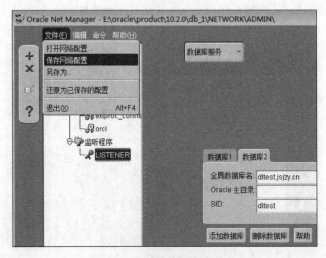

图 6-25 监听配置完成界面

实验 6.2 PowerDesigner 使用简介

在开发数据库管理信息系统时，设计数据库的数据表并建立它们之间的联系是比较烦琐的工作，可以借助一些工具来简化操作，本节介绍一款比较实用的数据库后台设计工具 PowerDesigner。

PowerDesigner 是 Sybase 公司的 CASE 工具集，使用它可以方便地对管理信息系统进行分析设计，它的功能涵盖了数据库模型设计的全过程。利用 PowerDesigner 可以制作数据流程图、概念数据模型、物理数据模型，还可以为数据仓库制作结构模型，也能对团队设计模型进行控制。它可以与许多流行的软件开发工具如 PowerBuilder、Delphi、VB 等相配合，使开发时间缩短和使系统设计更优化。

PowerDesigner 是进行数据库设计的强大软件，是一款开发人员常用的数据库建模工具。使用它可以分别从概念数据模型(Conceptual Data Model)和物理数据模型(Physical Data Model)两个层次对数据库进行设计。PowerDesigner 自带的各种可视化设计工具可以直观地建立所需的数据表，并描述它们之间的联系，最终还可以把它们转换为被 Oracle 或其他数据库管理系统识别的 SQL 脚本，从而便捷地完成数据库后台的建设。使用 PowerDesigner 会减少程序设计人员的工作量，并提高软件项目的开发效率。

【实验目的】

(1) 掌握安装 PowerDesigner 的方法。
(2) 掌握使用 PowerDesigner 建立概念模型的方法，并生成 SQL 语句。

【实验内容】

(1) 安装 PowerDesigner。
(2) 建立概念模型、建立物理模型。

【实验步骤】

1. 安装 PowerDesigner

运行 PowerDesigner 安装包，出现图 6-26 所示界面，单击 Install 按钮进入下一步。

在弹出的欢迎界面中单击 Next 按钮进入下一步，如图 6-27 所示。

在弹出的安装协议界面中选择 Peoples Republic of China(PRC) 选项，选择 I AGREE to the terms of the Sybase license, for the install location specified 单选按钮。单击 Next 按钮进入下一步，如图 6-28 所示。

在弹出的安装路径界面中单击 Browse 按钮选择安装的路径，这里选择安装在 D:\Program Files 路径。单击 Next 按钮进入下一步，如图 6-29 所示。

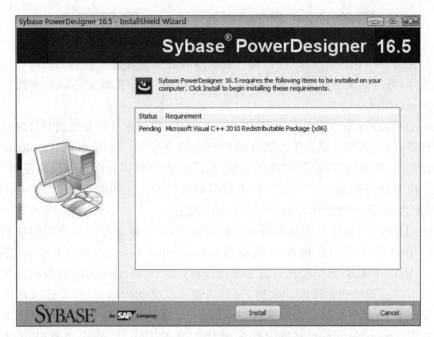

图 6-26　PowerDesigner 安装界面

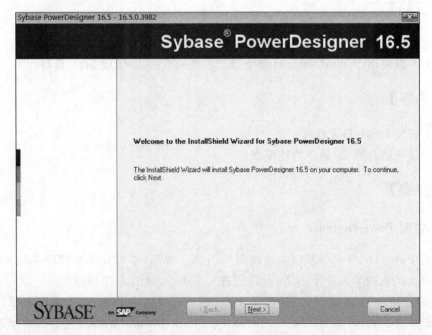

图 6-27　PowerDesigner 欢迎界面

第 6 章 手工建立数据库

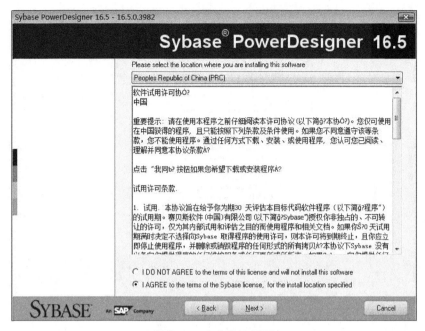

图 6-28 安装协议界面

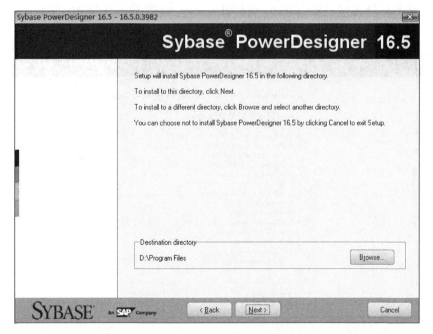

图 6-29 选择安装路径界面

在弹出的安装组件界面中选择需要安装的组件,可根据需求选择,这里选择系统默认的组件。单击 Next 按钮进入下一步,如图 6-30 所示。

在弹出的安装属性文件界面中选择需要安装的属性文件,可根据需求选择,这里选

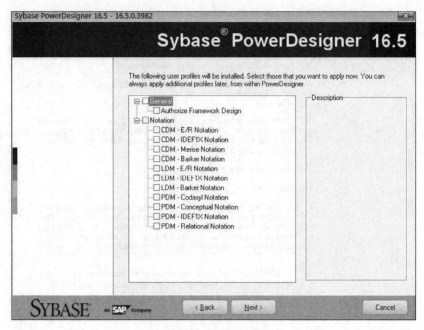

图 6-30　安装组件界面

择系统默认的属性文件。单击 Next 按钮进入下一步,如图 6-31 所示。

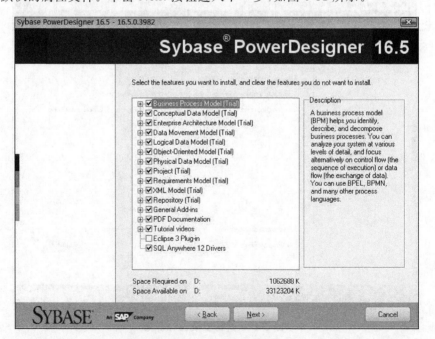

图 6-31　安装属性界面

在弹出的安装界面中选择"开始"菜单的显示名称。单击 Next 按钮进入下一步,如图 6-32 所示。

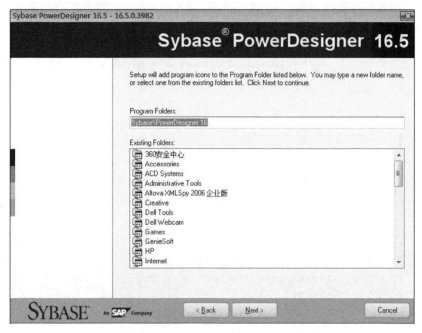

图 6-32　设置"开始"菜单的显示名称界面

在弹出的确认安装信息界面中确认安装信息。单击 next 按钮进入下一步,如图 6-33 所示。安装过程如图 6-34 所示,安装完成的界面如图 6-35 所示。

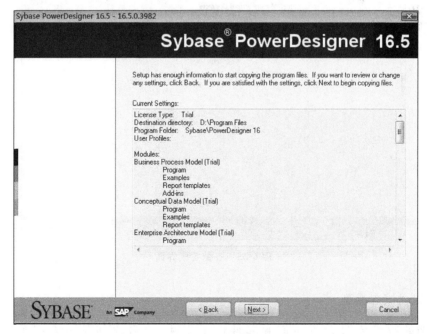

图 6-33　安装信息确认界面

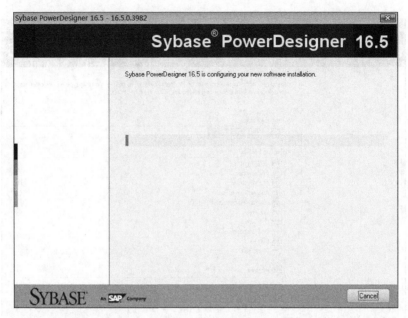

图 6-34 安装过程界面

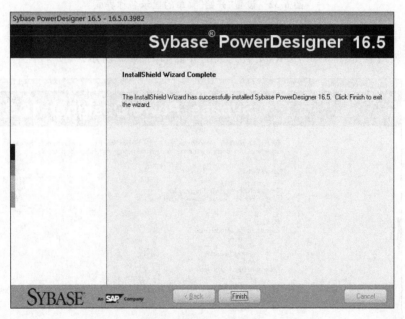

图 6-35 安装完成界面

2. 建立概念模型

本例的数据库包含供应商和零件两个实体,二者之间是多对多关系。使用 PowerDesigner 建立概念模型和逻辑模型。

打开 PowerDesigner 系统的主界面,选择 File→New Model 命令,在 New Model 对

话框中的 Model type 列表框中选择 Conceptual Data Model 类型,并在 Model name 文本框中为项目命名,本例为 ConceptualDataModel_sp,然后单击 OK 按钮,如图 6-36 所示。

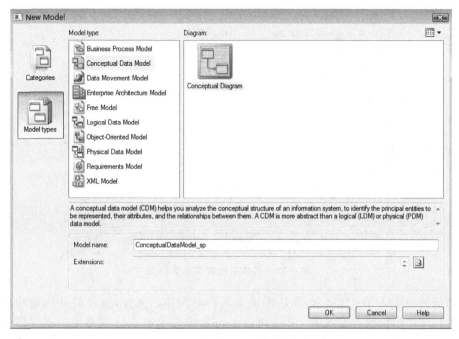

图 6-36　Conceptual Data Model 界面

从右侧的模板中单击实体 Entity 图标,并单击设计区域,将实体添加到设计视图中,重复此操作,共向设计视图中添加两个实体。再从右侧的模板中单击"联系"图标,并单击设计区域,将联系添加到设计视图中,向设计视图中添加一个联系,如图 6-37 所示。

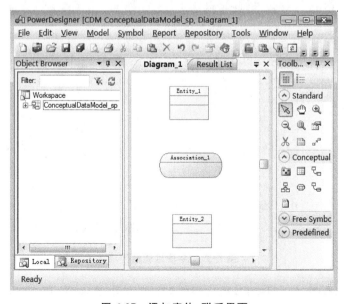

图 6-37　添加实体、联系界面

依次双击每一个实体和联系，在弹出的界面中选择 General 选项卡，并在 Name 文本框中输入对应的实体名。两个实体名分别为 supplier 和 part，如图 6-38 所示。

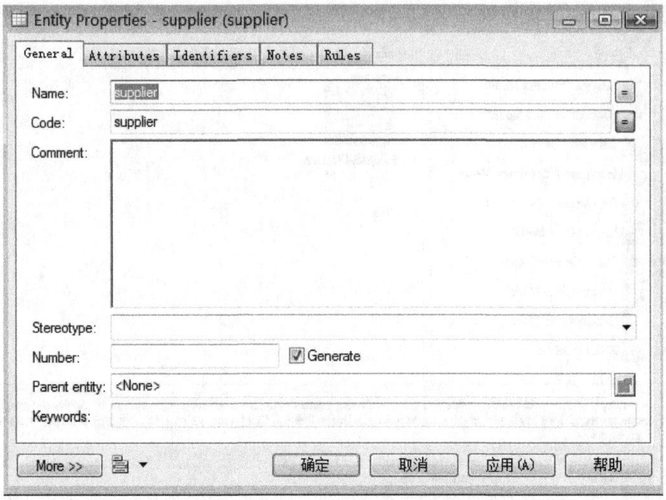

图 6-38　实体常规选项设置界面

切换到 Attributes 选项卡，在这里定义每一个实体的属性，也就是数据表中的每一个字段。字段编辑器共有 5 列，第一列是 Name（字段名），第二列是 Code（字段的标识符），第三列是 Data Type（字段的数据类型），第四列为 Length（数据长度），第五列为 Precision（数据精度）。后面有三个复选框，分别为 M、P、D，选中 M 表示此字段不能为空，选中 P 表示此字段为主键，选中 D 表示此字段的信息将显示在设计视图中。字段编辑器如图 6-39 所示，属性设计结果如图 6-40 所示。

图 6-39　字段编辑器界面

设计实体和联系之间的关系，从右侧的模板中单击 Association link 图标，在需要建立关系的实体和联系之间拖曳，建立这两者之间的关系，如图 6-41 所示。

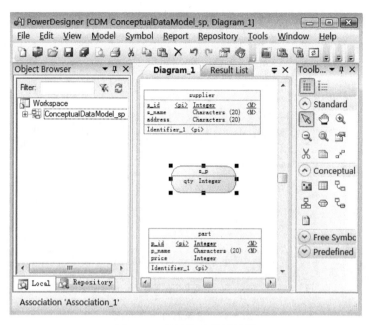

图 6-40 属性设计结果界面

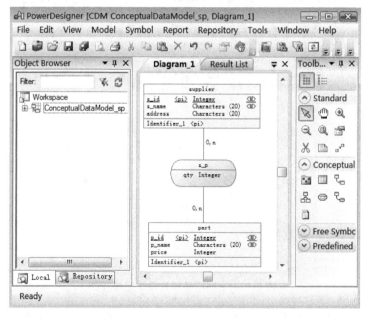

图 6-41 建立实体之间联系界面

依次双击每一个联系,在弹出的 Association Link Properties 窗口中选择 General 选项卡,并在 Entity 和 Association 文本框中输入对应的名称,本例为 supplier 和 s_p。在 Cardinality 下拉列表框中选择联系种类,本例为 1∶n。单击"确定"按钮,如图 6-42 所示。其他联系属性设计完毕的界面如图 6-43 所示。

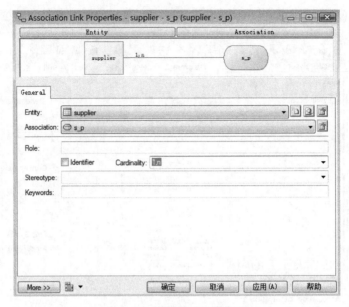

图 6-42 联系属性设置界面

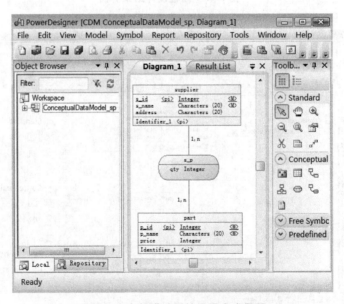

图 6-43 全部联系属性设置完成界面

至此,关系的概念模型已全部设计完成,接下来生成物理数据模型。选择 Tools→Generate Physical Data Model 命令,弹出 PDM Generation Options 对话框,如图 6-44 所示。

在 General 选项卡中选择 Generate new Physical Data Model 单选按钮,并选择 DBMS 为 ORACLE Version 10g,Name 为 ConceptualDataModel_sp,单击"确定"按钮。在弹出的路径选择对话框中输入要存放生成的物理模型文件的位置,系统会在该路径下

图 6-44 物理数据模型设置界面

自动生成一个名为 ConceptualDataModel_sp.pdm 的文件,并生成系统的物理模型,如图 6-45 所示。

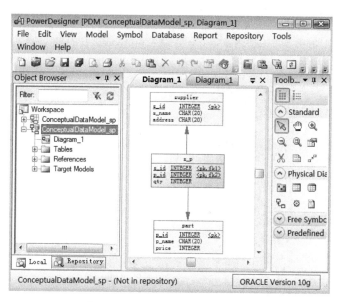

图 6-45 生成物理数据模型界面

选择 Database→Generat Database 命令,将生成的物理模型转换为 SQL 语句,以供 Oracle 数据库管理系统编译,并生成数据库实体,如图 6-46 所示。在 Directory 下拉列表框中输入生成文件的存放路径,并在 File name 下拉列表框中输入生成文件的文件名,单击"确定"按钮。PowerDesigner 将会在指定路径下生成一个名为.sql 的文本文件,里面存放可供 Oracle 执行的用来建立数据库实体的 SQL 语句。

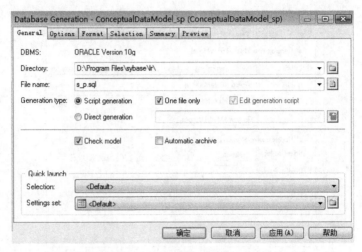

图 6-46　SQL 语句文件生成界面

第 7 章 项目实践

项目实践对于巩固数据库知识,加强实际动手能力和提高综合素质十分必要。通过项目实践,培养 C/S 和 B/S/模式的数据库应用软件系统的设计和开发能力,进一步熟悉数据库管理系统的操作技术,提高分析问题和解决问题的能力。

(1) 熟练掌握数据库管理系统 Oracle 的使用。
(2) 熟练掌握一种数据库应用软件开发工具(如 Java、ASP、VB. NET)的使用。

项目 7.1 考试报名管理系统

【课程设计任务与要求】

1. 任务

(1) 考试报名系统中有各种科目的考试及竞赛类活动,还有报考学生的详细信息。
(2) 每种考试都有考试名称、考试代号、考试时间、考试地点。
(3) 每个考生有考号、姓名、专业、学校、性别、年龄、登录密码。
(4) 每个学生可报考多门考试或竞赛,每个考试或竞赛可供多个学生选择。
(5) 管理员可对所有信息进行插入、查询、修改、删除等操作,报考者只可以进行查询操作。

2. 设计要求

(1) 实现新考试科目或竞赛的数据录入。
(2) 实现所有考试科目或竞赛的数据修改和查询。
(3) 实现对所有考生信息的数据录入、查询、更新、删除。
(4) 能够进行考生报考信息的管理。
(5) 能够进行用户管理。

【课程设计步骤】

1. 需求分析

1) 数据需求

考试报名管理信息系统需要完成的功能主要有：

（1）考试信息的更新、查询，包括考试名称、考试代号、考试时间、考试地点。

（2）报考者基本信息的查询、更新，包括考号、姓名、专业、学校、性别、年龄、登录密码。

（3）新报考者基本信息的插入，包括考号、姓名、专业、学校、性别、年龄、登录密码。

（4）新考试信息的输入，包括考试名称、考试代号、考试时间、考试地点。

（5）报考信息的插入和查询，包括考生的考号、考试代号。

（6）管理员对普通用户的添加、删除。

2) 事务需求

（1）在考生信息管理部分要求：

① 可以查询考生信息。

② 可以对考生信息进行添加、更新操作。

（2）考试信息管理部分要求：

① 可以查询考试信息。

② 可以对考试信息进行维护，包括添加及更新操作。

（3）报考信息管理部分要求：

① 可以查询报考信息。

② 可以对报考信息进行维护操作。

（4）用户信息管理部分要求：

① 可以插入用户信息。

② 可以对用户信息进行删除操作。

2. 概要设计

根据所要实现的功能设计，建立实体之间的关系，进而实现逻辑结构功能。

考试报名管理信息系统可以划分为考生信息实体、考试信息实体和管理员实体。系统 E-R 图如图 7-1 所示。

3. 逻辑设计

根据图 7-1 建立系统的基本表，如表 7-1～表 7-4 所示。

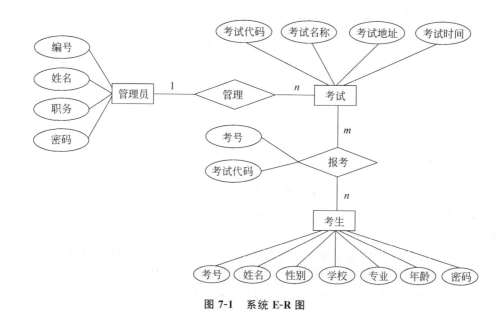

图 7-1　系统 E-R 图

表 7-1　考试信息表

表中列名	数据类型	可否为空	说明
address	Char(30)	not null	地址
num	Char(30)	not null(主键)	代码
time	Char(30)	not null	时间
ename	Char(20)	not null	名称

表 7-2　考生信息表

表中列名	数据类型	可否为空	说明
Snum	Char(20)	not null(主键)	考号
sname	Char(20)	not null	姓名
age	Int(4)	not null	年龄
school	Char(20)	not null	学校
sex	Char(20)	not null	性别
major	Char(20)	not null	专业
password	Char(20)	not null	密码

表 7-3　报考信息表

表中列名	数据类型	可否为空	说明
snum	Char(20)	Not null(主键)	考号
num	Char(20)	Not null(主键)	考试代码

表 7-4 管理员信息表

表中列名	数据类型	可否为空	说明
loginnum	Char(20)	not null（主键）	编号
name	Char(20)	not null	姓名
caree	Char(20)	not null	职务
password	Char(20)	not null	密码

4．物理设计

数据库物理设计阶段的任务是根据具体计算机系统（DBMS 和硬件等）的特点，为给定的数据库模型确定合理的存储结构和存取方法。所谓的"合理"主要有两个含义：一是要使设计出的物理数据库占用较少的存储空间；二是对数据库的操作具有尽可能高的速度。确定数据库的存储结构主要指确定数据的存放位置和存储结构，包括确定关系、索引、日志、备份等的存储安排及存储结构，以及确定系统存储参数的配置。

（1）对考试信息表（exam）在 num 属性列上建立主键。

（2）对学生表（student）在 snum 上建立主键。

（3）对管理员表（superuser）在 loginnum 上建立主键。

（4）对报考表（entryform）在 snum、num 上建立主键。

5．数据库建立

（1）创建考试表。

```
CREATE TABLE 'exam' (
'address' CHAR(50) NOT NULL,
'num' CHAR(50) NOT NULL,
'time' CHAR(50) NOT NULL,
'ename' CHAR(50) NOT NULL,
PRIMARY KEY ('num'));
```

（2）创建学生表。

```
CREATE TABLE 'student' (
'snum' CHAR(50) NOT NULL,
'sname' CHAR(50) NOT NULL,
'age' INT(4) NOT NULL,
'school' CHAR(50) NOT NULL,
'sex' CHAR(50) NOT NULL,
'major' CHAR(50) NOT NULL,
'password' CHAR(50) NOT NULL,
PRIMARY KEY ('snum'));
```

（3）创建报考表。

```
CREATE TABLE 'entryform' (
```

```
'snum' CHAR(50) NOT NULL,
'num' CHAR(50) NOT NULL,
PRIMARY KEY ('snum', 'num')
);
```

(4)创建管理员表。

```
CREATE TABLE 'superuser' (
'loginnum' CHAR(50) NOT NULL,
'name' CHAR(50) NOT NULL,
'caree' CHAR(50) NOT NULL,
'password' CHAR(50) NOT NULL,
PRIMARY KEY ('loginnum'));
```

(5)数据库用户权限管理。

该系统设置三种类型的用户:

- 超级管理员(Manager):即系统管理员拥有所有的权限。
- 管理员:可以进行查询。
- 考生:只能查询信息。

6. 建立触发器

(1)建立学生表增加触发器。

目的:将 student 表中添加的学生信息自动备份到 student0 中,防止 student 表损坏后数据丢失。

```
CREATE TRIGGER 'a' AFTER INSERT ON 'student' FOR EACH ROW BEGIN
insert into student0 values(new.snum,new.sname,new.age,new.school,new.sex,
new.major,new.password);
END;
```

(2)建立学生表修改触发器。

目的:当 student 表中有学生信息修改时,自动将被修改过的信息行复制一份插入到 studentUpdate2 中。

```
CREATE TRIGGER 'u' AFTER UPDATE ON 'student' FOR EACH ROW BEGIN
update studentupdate set new.sname=now(),new.age=now(),new.school=now(),
new.sex=now(),new.major=now(),new.password=now() where student.snum=
studentupdate.snum;
END;
```

7. 系统功能模块

系统功能模块图如图 7-2 所示。

5 个子系统的功能如下:

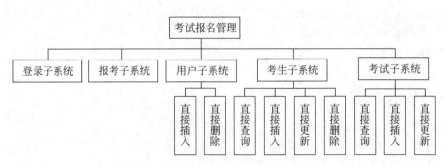

图 7-2 系统功能模块图

(1) 登录子系统：可以实现管理员和用户登录管理。
(2) 报考子系统：主要用于用户查询和添加报考信息。
(3) 用户子系统：主要用于管理员对用户进行插入和删除。
(4) 考生子系统：主要用于考生查询、更新信息的操作。
(5) 考试子系统：主要用于管理员对考试信息的查询、更新。

项目 7.2 图书管理系统

【课程设计任务与要求】

1. 任务

(1) 每种图书有书号、书名、出版社、作者、库存量。
(2) 借书记录有学生的学号、书号、登记时期等。

2. 设计要求

(1) 实现新进图书的数据录入和下架图书的数据删除。
(2) 实现借阅者图书证信息的录入和删除。
(3) 能够记录学生的个人资料和所借图书的书名、书号、借书时间等。
(4) 能够进行借书还书处理。
(5) 能够进行用户管理。

【课程设计步骤】

1. 需求分析

1) 数据需求
(1) 读者基本信息的输入，包括借书证编号、读者姓名、读者性别等。
(2) 读者基本信息的查询、修改，包括读者借书证编号、读者姓名、读者性别等。
(3) 书籍库存信息的输入，包括书籍编号、书籍名称、书籍类别、作者姓名、出版社名称、出版日期、登记日期。

(4) 书籍库存信息的查询、修改,包括书籍编号、书籍名称、书籍类别、作者姓名、出版社名称、出版日期登记日期等。

(5) 借书信息的输入,包括读者借书证编号、书籍编号、借书日期。

(6) 借书信息的查询、修改,包括读者借书证编号、读者姓名、书籍编号、书籍名称、借书日期等。

(7) 还书信息的输入,包括借书证编号、书籍编号、还书日期。

(8) 还书信息的查询和修改,包括还书读者借书证编号、读者姓名、书籍编号、书籍名称、借书日期、还书日期等。

(9) 管理员管理,包括创建读者用户信息、删除读者用户信息、添加图书信息、删除图书信息。

2) 事务需求

(1) 读者信息管理部分要求:

① 可以查询读者信息。

② 可以对读者信息进行添加及删除操作。

(2) 书籍信息管理部分要求:

① 可以浏览书籍信息。

② 可以对书籍信息进行维护,包括添加及删除操作。

(3) 借阅信息管理部分要求:

① 可以浏览借阅信息。

② 可以对借阅信息进行维护操作。

(4) 归还信息管理部分要求:

① 可以浏览归还信息。

② 可以对归还信息进行修改维护操作。

(5) 管理者信息管理部分要求:

① 显示当前数据库中管理者情况。

② 对管理者信息维护操作。

2. 概要设计

根据所要实现的功能设计,建立实体之间的关系,进而实现逻辑结构。

考试报名管理信息系统可以划分为学生、图书、借阅信、借阅管理、图书管理、普通管理员、超级管理员实体。系统 E-R 图如图 7-3 所示。

3. 逻辑设计

根据图 7-3 建立系统的基本表,如表 7-5~表 7-11 所示。

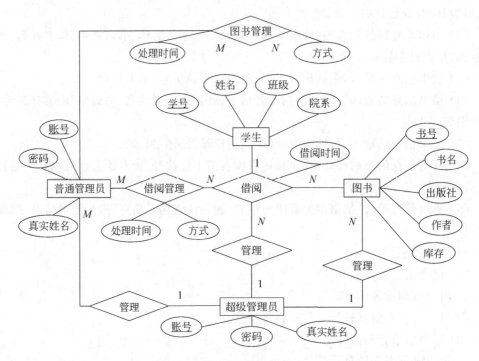

图 7-3 系统 E-R 图

表 7-5 学生表

表中列名	数据类型	可否为空	说　明
sno	varchar(10)	not null(主键)	学号
sna	varchar(8)	not null	学生姓名
sex	varchar(2)	not null	学生性别
cla	varchar(10)	not null	班级
aca	varchar(18)	not null	系别

表 7-6 图书表

表中列名	数据类型	可否为空	说　明
bno	Varchar(4)	Not null(主键)	书号
bna	Varchar(16)	Not null	书名
pre	Varchar(16)	Not null	出版社
wri	Varchar(16)	null	作者
sum	int	Check sum>=0	库存

表 7-7 借阅信息表

表中列名	数据类型	可否为空	说　明
sno	Varchar(10)	Not null(外主键)	学号
bno	Varchar(4)	Not null(外主键)	书号
Bor_time	Varchar(10)	Not null	借书时间

表 7-8　借阅管理表

表中列名	数据类型	可否为空	说　明
Adacc	Varchar(10)	Not null	管理员账号
bno	Varchar(4)	Not null	书号
sno	Varchar(10)	Not null	学号
time	Varchar(20)	Not null（主键）	时间
way	Varcahr(4)	Check	方法（借出/归还）

表 7-9　图书管理表

表中列名	数据类型	可否为空	说　明
Adacc	Varchar(10)	Not null	管理员账号
Bno	Varchar(4)	Not null	书号
Bna	Varchar(16)	Not null	书名
Pre	Varchar(16)	Not null	出版社
Wri	Varchar(16)	Null	作者
Time	Varchar(20)	Not null（主键）	时间
way	Varchar(4)	check	方法（添加/删除）

表 7-10　普通管理员

表中列名	数据类型	可否为空	说　明
Adacc	varchar(10)	not null（主键）	管理员账号
Adpsd	varchar(16)	not null	管理员密码
adna	varchar(10)	not null	管理员真实姓名

表 7-11　超级管理员表

表中列名	数据类型	可否为空	说　明
suacc	varchar(10)	not null（主键）	超级管理员号
supsd	varchar(16)	not null	超级管理员姓名
suna	varchar(10)	not null	超级管理员真实姓名

4．物理设计

数据库物理设计阶段的任务是根据具体计算机系统（DBMS 和硬件等）的特点，为给定的数据库模型确定合理的存储结构和存取方法。所谓"合理"主要有两个含义：一是要使设计出的物理数据库占用较少的存储空间；二是对数据库的操作具有尽可能高的速度。

确定数据库的存储结构主要是指确定数据的存放位置和存储结构，包括确定关系、索引、日志、备份等的存储安排及存储结构，以及确定系统存储参数的配置。

建立索引：

//学生表（班级列）建立索引
CREATE INDEX S_I_CLA ON STUDENT(CLA);

```
//图书表(作者列)建立索引
CREATE INDEX B_I_WRI ON BOOK(WRI);
//图书表(出版社列)建立索引
CREATE INDEX B_I_PRE ON BOOK(PRE);
//普通管理员表(真实姓名列)建立索引
CREATE INDEX AD_I_ADNA ON SYS_AD(ADNA);
```

5. 数据库建立

(1)学生表：(<u>学号</u>,姓名,性别,班级,学院)

```
CREATE TABLE STUDENT
(
    SNO VARCHAR(10) NOT NULL,
    SNA VARCHAR(8)  NOT NULL,
    SEX VARCHAR(2)  NOT NULL,
    CLA VARCHAR(10) NOT NULL,
    ACA VARCHAR(18) NOT NULL,
    PRIMARY KEY(SNO)
);
```

(2)图书表：(<u>书号</u>,书名,出版社,作者,库存)

```
//pre 出版社 WRI 作者 sum 库存
CREATE TABLE BOOK
(
    BNO VARCHAR(4) NOT NULL,
    BNA VARCHAR(16)NOT NULL,
    PRE VARCHAR(16)NOT NULL,
    WRI VARCHAR(16),
    SUM INT CHECK(SUM>=0),
    PRIMARY KEY(BNO)
);
```

(3)借阅信息表：(<u>学号,书号</u>,借阅时间)

```
CREATE TABLE BOR_BOOK
(
    SNO VARCHAR(10) NOT NULL,
    BNO VARCHAR(4)  NOT NULL,
    BOR_TIME varchar(10) NOT NULL,
    PRIMARY KEY(SNO,BNO),
    FOREIGN KEY(SNO) REFERENCES STUDENT(SNO),
    FOREIGN KEY(BNO) REFERENCES BOOK(BNO)
);
```

(4) 借阅管理表:(管理员账号,书号,学号,处理时间,方式(借出/归还))

```
CREATE TABLE MA_BOR
(
    ADACC VARCHAR(10) NOT NULL,
    BNO VARCHAR(4)    NOT NULL,
    SNO VARCHAR(10)   NOT NULL,
    TIME VARCHAR(20)  NOT NULL,
    WAY VARCHAR(4) CHECK(WAY IN('借出','归还')),
    PRIMARY KEY(TIME)
);
```

(5) 图书管理表:(管理员账号,书号,书名,出版社,作者,处理时间,方式(添加/删除))

```
CREATE TABLE MA_BOOK
(
    ADACC VARCHAR(10) NOT NULL,
    BNO VARCHAR(4)    NOT NULL,
    BNA VARCHAR(16)   NOT NULL,
    PRE VARCHAR(16)   NOT NULL,
    WRI VARCHAR(16),
    TIME VARCHAR(20)  NOT NULL,
    WAY VARCHAR(4) CHECK(WAY IN('添加','删除')),
    PRIMARY KEY(TIME)
);
```

(6) (普通)管理员:(管理员账号,密码,真实姓名)

```
CREATE TABLE SYS_AD
(
    ADACC VARCHAR(10) NOT NULL,
    ADPSD VARCHAR(16) NOT NULL,
    ADNA  VARCHAR(10),
    PRIMARY KEY (ADACC)
);
```

(7) 超级管理员:(管理员账号,密码,真实姓名)

```
CREATE TABLE SYS_SU_AD
(
    SUACC VARCHAR(10) NOT NULL,
    SUPSD VARCHAR(16) NOT NULL,
    SUNA  VARCHAR(10),
    PRIMARY KEY (SUACC)
);
```

6. 数据库用户权限管理

该系统设置两种类型的用户：

（1）超级管理员（Superadministrator）。即系统管理员拥有所有的权限，同时具有管理普通管理员的权限。

（2）管理员（Administrator）。可以进行借还书处理，添加和删除图书。

7. 建立触发器

1）图书库存自减触发器

功能：当有人借书后（即往借阅信息中插入数据），使图书表中对应书号的库存字段减1，图书表中对应库存相应字段用 check 进行约束，使其值最小为0，不会出现负值。

```
CREATE TRIGGER TR_SUB_SUM
AFTER INSERT ON BOR_BOOK
for each row
BEGIN
    UPDATE BOOK SET SUM=SUM-1
    WHERE BNO=:new.bno;

END;
```

2）图书库存自增触发器

功能：当有人还书后（即从借阅信息中删除数据），使图书表中对应书号的库存字段加1。

```
CREATE TRIGGER TR_ADD_SUM
AFTER DELETE ON BOR_BOOK
FOR EACH ROW
BEGIN
    UPDATE BOOK SET SUM=SUM+1
    WHERE BNO=:OLD.BNO;
END;
```

3）借阅管理数据更新触发器——"借书"

功能：当有人借书后，在借阅管理表中自动记录（管理员账号，书号，学号，执行时间，管理方式（借出））。

```
CREATE TRIGGER TR_MA_BOR1
AFTER INSERT ON BOR_BOOK
FOR EACH ROW
DECLARE
    PADACC VARCHAR(10);
    PWAY VARCHAR(4);
    PTIME VARCHAR(20);
```

```
BEGIN
    select user INTO PADACC from dual;
    select to_char(sysdate, 'yyyy/MM/dd HH24:MI:SS' ) INTO PTIME from dual;
    PWAY:='借出';
    INSERT INTO MA_BOR(ADACC,BNO,SNO,TIME,WAY)
    VALUES(PADACC,:NEW.BNO,:NEW.SNO,PTIME,PWAY);

END;
```

4) 借阅管理数据更新触发器——"还书"

功能：当有人还书后，在借阅管理表中自动记录(管理员账号，书号，学号，执行时间，管理方式(归还))。

```
CREATE TRIGGER TR_MA_BOR2
AFTER DELETE ON BOR_BOOK
FOR EACH ROW
DECLARE
    PADACC VARCHAR(10);
    PWAY VARCHAR(4);
    PTIME VARCHAR(20);

BEGIN
    select user INTO PADACC from dual;
    select to_char(sysdate, 'yyyy/MM/dd HH24:MI:SS' ) INTO PTIME from dual;
    PWAY:='归还';
    INSERT INTO MA_BOR(ADACC,BNO,SNO,TIME,WAY)
    VALUES(PADACC,:OLD.BNO,:OLD.SNO,PTIME,PWAY);
END;
```

5) 图书管理数据更新触发器——"添加"

功能：当管理员在图书表中添加图书后，在图书管理表中自动记录(管理员账号，书号，书名，出版社，作者，处理时间，方式(添加))。

```
CREATE TRIGGER TR_MA_BOOK1
AFTER INSERT ON BOOK
FOR EACH ROW
DECLARE
    PADACC VARCHAR(10);
    PWAY VARCHAR(4);
    PTIME VARCHAR(20);

BEGIN
    select user INTO PADACC from dual;
    select to_char(sysdate, 'yyyy/MM/dd HH24:MI:SS' ) INTO PTIME from dual;
    PWAY:='添加';
```

```
        INSERT INTO MA_BOOK(ADACC,BNO,BNA,PRE,WRI,TIME,WAY)
        VALUES(PADACC,:NEW.BNO,:NEW.BNA,:NEW.PRE,:NEW.WRI,PTIME,PWAY);

END;
```

6）图书管理数据更新触发器——"删除"

功能：当管理员在图书表中删除图书后，在图书管理表中自动记录（管理员账号，书号，书名，出版社，作者，处理时间，方式（删除））。

```
CREATE TRIGGER TR_MA_BOOK2
AFTER DELETE ON BOOK
FOR EACH ROW
DECLARE
    PADACC VARCHAR(10);
    PWAY VARCHAR(4);
    PTIME VARCHAR(20);

BEGIN
    select user INTO PADACC from dual;
    select to_char(sysdate, 'yyyy/MM/dd HH24:MI:SS' ) INTO PTIME from dual;
    PWAY:='删除';
    INSERT INTO MA_BOOK(ADACC,BNO,BNA,PRE,WRI,TIME,WAY)
        VALUES(PADACC,:OLD.BNO,:OLD.BNA,:OLD.PRE,:OLD.WRI,PTIME,PWAY);

END;
```

8. 系统功能模块

三个子系统的功能如下：

（1）登录子系统：可以实现图书管理员和超级管理员登录管理。

（2）借还子系统：主要由普通管理员进行借书还书记录的登记和清除。

（3）管理子系统：由普通管理员和超级管理员管理系统，其中普通管理员或超级管理员管理图书（包括图书信息的修改、新图书的增加、旧图书的删除），借阅信息（借书时添加信息、归还时删除信息）；超级管理员创建和删除普通管理员。

附录

SQL * Plus 环境命令

Oracle 的 SQL * Plus 是与 Oracle 进行交互的客户端工具。在 SQL * Plus 中可以运行 SQL * Plus 环境命令与 SQL * Plus 语句。

通常所说的 DML、DDL、DCL 语句都是 SQL * Plus 语句,它们执行完后都可以保存在一个被称为 SQL Buffer 的内存区域中,并且只能保存一条最近执行的 SQL 语句。可以对保存在 SQL Buffer 中的 SQL 语句进行修改,然后再次执行。SQL 语句一般都与数据库打交道。

除了 SQL * Plus 语句外,在 SQL * Plus 中执行的其他语句称为 SQL * Plus 环境命令。它们执行完后不保存在 SQL Buffer 的内存区域中,一般用来对输出的结果进行格式化显示,以便于制作报表。

下面就介绍一些常用的 SQL * Plus 命令。

1. 执行一个 SQL 脚本文件

```
SQL>start file_name
SQL>@file_name
```

可以将多条 SQL 语句保存在一个文本文件中,这样当要执行这个文件中的所有 SQL 语句时,用上面的任一命令即可,这类似于 DOS 中的批处理。

@与@@的区别是什么?

@等于 start 命令,用来运行一个 SQL 脚本文件。

@命令调用当前目录下的,或指定全路径的脚本文件。该命令一般指定要执行的文件的全路径,否则从默认路径下读取指定的文件。

@@用在 SQL 脚本文件中,用来说明用@@执行的 SQL 脚本文件与@@所在的文件在同一目录下,而不用指定要执行 SQL 脚本文件的全路径,也不是从 SQLPATH 环境变量指定的路径中寻找 SQL 脚本文件。该命令一般用在脚本文件中。

如在 c:\temp 目录下有文件 start.sql 和 nest_start.sql,start.sql 脚本文件的内容为 @@nest_start.sql 相当于@ c:\temp\nest_start.sql,则在 SQL * Plus 中这样执行:

```
SQL>@c:\temp\start.sql
```

2. 对当前的输入进行编辑

```
SQL>edit
```

3. 重新运行上一次运行的 SQL 语句

```
SQL>/
```

4. 将显示的内容输出到指定文件

```
SQL>SPOOL file_name
```

在屏幕上的所有内容都包含在该文件中,包括输入的 SQL 语句。

5. 关闭 spool 输出

```
SQL>spool off
```

只有关闭 spool 输出才会在输出文件中看到输出的内容。

6. 显示一个表的结构

```
SQL>desc table_name
```

7. COL 命令

主要格式化列的显示形式。
该命令有许多选项,具体如下:

```
COL[UMN] [{ column|expr} [ option ...]]
```

Option 选项可以是如下子句:

```
ALI[AS] alias
CLE[AR]
FOLD_A[FTER]
FOLD_B[EFORE]
FOR[MAT] format
HEA[DING] text
JUS[TIFY] {L[EFT]|C[ENTER]|C[ENTRE]|R[IGHT]}
LIKE { expr|alias }
NEWL[INE]
NEW_V[ALUE] variable
NOPRI[NT]|PRI[NT]
NUL[L] text
OLD_V[ALUE] variable
ON|OFF
```

WRA[PPED]|WOR[D_WRAPPED]|TRU[NCATED]

(1) 改变默认的列标题。

COLUMN column_name HEADING column_heading
```
SQL>select * from dept;
DEPTNO DNAME                          LOC
------ ---------------------------    ---------
    10 ACCOUNTING                     NEW YORK
SQL>col LOC heading location
SQL>select * from dept;
DEPTNO DNAME                          location
------ ---------------------------    -----------
    10 ACCOUNTING                     NEW YORK
```

(2) 将列名 ENAME 改为新列名 EMPLOYEE NAME,并将新列名放在两行上。

```
SQL>select * from emp
Department   name       Salary
----------   ---------- ----------
        10   aaa        11
SQL>COLUMN ENAME HEADING 'Employee|Name'
Sql>select * from emp
             Employee
Department   name       Salary
----------   ---------- ----------
        10   aaa        11
```
note: the col heading turn into two lines from one line.

(3) 改变列的显示长度。

FOR[MAT] format
```
SQL>select empno,ename,job from emp;
EMPNO  ENAME       JOB
------ ----------  ---------
  7369 SMITH       CLERK
  7499 ALLEN       SALESMAN
  7521 WARD        SALESMAN
SQL>col ename format a40
EMPNO  ENAME       JOB
------ ----------- ---------
  7369 SMITH       CLERK
  7499 ALLEN       SALESMAN
  7521 WARD        SALESMAN
```

(4) 设置列标题的对齐方式。

```
JUS[TIFY] {L[EFT]|C[ENTER]|C[ENTRE]|R[IGHT]}
SQL>col ename justify center
SQL>/
EMPNO   ENAME         JOB
------  ------------  ---------
  7369    SMITH       CLERK
  7499    ALLEN       SALESMAN
  7521    WARD        SALESMAN
```

对于 NUMBER 型的列，列标题默认在右边，其他类型的列标题默认在左边。

(5) 不让一个列显示在屏幕上。

```
NOPRI[NT]|PRI[NT]
SQL>col job noprint
SQL>/
EMPNO   ENAME
------  ------------
  7369    SMITH
  7499    ALLEN
  7521    WARD
```

(6) 格式化 NUMBER 类型列的显示。

```
SQL>COLUMN SAL FORMAT $99,990
SQL>/
Employee
Department  Name     Salary     Commission
----------  -------  ---------  ----------
30          ALLEN    $1,600     300
```

(7) 显示列值时，如果列值为 NULL 值，用 text 值代替 NULL 值。

```
COMM NUL[L] text
SQL>COL COMM NUL[L] text
```

(8) 设置一个列的回绕方式。

```
WRA[PPED]|WOR[D_WRAPPED]|TRU[NCATED]
    COL1
--------------------
HOW ARE YOU?


SQL>COL COL1 FORMAT A5
SQL>COL COL1 WRAPPED
COL1
-----
```

```
HOW A
RE YO
U?

SQL>COL COL1 WORD_WRAPPED
COL1
-----
HOW
ARE
YOU?
```

（9）显示列当前的显示属性值。

```
SQL>COLUMN column_name
```

（10）将所有列的显示属性设为默认值。

```
SQL>CLEAR COLUMNS
```

8. 屏蔽掉一个列中显示的相同值

```
BREAK ON break_column
SQL>BREAK ON DEPTNO
SQL>SELECT DEPTNO, ENAME, SAL
     FROM EMP
       WHERE SAL <2500
         ORDER BY DEPTNO;
DEPTNO    ENAME       SAL
--------- ----------- ---------
10        CLARK       2450
          MILLER      1300
20        SMITH       800
          ADAMS       1100
```

9. 在上面屏蔽掉一个列中显示的相同值的显示中，每当列值变化时在值变化之前插入 n 个空行

```
BREAK ON break_column SKIP n

SQL>BREAK ON DEPTNO SKIP 1
SQL>/
DEPTNO    ENAME    SAL
--------- -------- ---------
10 CLARK 2450
   MILLER 1300
```

```
20 SMITH 800
   ADAMS 1100
```

10. 显示对 BREAK 的设置

```
SQL>BREAK
```

11. 删除 8、9 的设置

```
SQL>CLEAR BREAKS
```

12. SET 命令

该命令包含许多子命令：

SET system_variable value

system_variable value 可以是如下子句之一：

```
APPI[NFO]{ON|OFF|text}
ARRAY[SIZE] {15|n}
AUTO[COMMIT]{ON|OFF|IMM[EDIATE]|n}
AUTOP[RINT] {ON|OFF}
AUTORECOVERY [ON|OFF]
AUTOT[RACE] {ON|OFF|TRACE[ONLY]} [EXP[LAIN]] [STAT[ISTICS]]
BLO[CKTERMINATOR] {.|c}
CMDS[EP] {;|c|ON|OFF}
COLSEP {_|text}
COM[PATIBILITY]{V7|V8|NATIVE}
CON[CAT] {.|c|ON|OFF}
COPYC[OMMIT] {0|n}
COPYTYPECHECK {ON|OFF}
DEF[INE] {&|c|ON|OFF}
DESCRIBE [DEPTH {1|n|ALL}] [LINENUM {ON|OFF}] [INDENT {ON|OFF}]
ECHO {ON|OFF}
EDITF[ILE] file_name[.ext]
EMB[EDDED] {ON|OFF}
ESC[APE] {\|c|ON|OFF}
FEED[BACK] {6|n|ON|OFF}
FLAGGER {OFF|ENTRY |INTERMED[IATE]|FULL}
FLU[SH] {ON|OFF}
HEA[DING] {ON|OFF}
HEADS[EP] {||c|ON|OFF}
INSTANCE [instance_path|LOCAL]
LIN[ESIZE] {80|n}
LOBOF[FSET] {n|1}
```

```
LOGSOURCE [pathname]
LONG {80|n}
LONGC[HUNKSIZE] {80|n}
MARK[UP] HTML [ON|OFF] [HEAD text] [BODY text] [ENTMAP {ON|OFF}] [SPOOL
{ON|OFF}] [PRE[FORMAT] {ON|OFF}]
NEWP[AGE] {1|n|NONE}
NULL text
NUMF[ORMAT] format
NUM[WIDTH] {10|n}
PAGES[IZE] {24|n}
PAU[SE] {ON|OFF|text}
RECSEP {WR[APPED]|EA[CH]|OFF}
RECSEPCHAR {_|c}
SERVEROUT[PUT] {ON|OFF} [SIZE n] [FOR[MAT] {WRA[PPED]|WOR[D_
WRAPPED]|TRU[NCATED]}]
SHIFT[INOUT] {VIS[IBLE]|INV[ISIBLE]}
SHOW[MODE] {ON|OFF}
SQLBL[ANKLINES] {ON|OFF}
SQLC[ASE] {MIX[ED]|LO[WER]|UP[PER]}
SQLCO[NTINUE] {>|text}
SQLN[UMBER] {ON|OFF}
SQLPRE[FIX] {#|c}
SQLP[ROMPT] {SQL>|text}
SQLT[ERMINATOR] {;|c|ON|OFF}
SUF[FIX] {SQL|text}
TAB {ON|OFF}
TERM[OUT] {ON|OFF}
TI[ME] {ON|OFF}
TIMI[NG] {ON|OFF}
TRIM[OUT] {ON|OFF}
TRIMS[POOL] {ON|OFF}
UND[ERLINE] {-|c|ON|OFF}
VER[IFY] {ON|OFF}
WRA[P] {ON|OFF}
```

(1) 设置当前 session 是否对修改的数据进行自动提交。

```
SQL>SET AUTO[COMMIT] {ON|OFF|IMM[EDIATE]|n}
```

(2) 在用 start 命令执行一个 SQL 脚本时,是否显示脚本中正在执行的 SQL 语句。

```
SQL>SET ECHO {ON|OFF}
```

(3) 是否显示当前 SQL 语句查询或修改的行数。

```
SQL>SET FEED[BACK] {6|n|ON|OFF}
```

默认只有结果大于 6 行时才显示结果的行数。如果 set feedback 1,则不管查询到多少行都返回。当为 off 时,一律不显示查询的行数。

(4) 是否显示列标题。

SQL>SET HEA[DING] {ON|OFF}

当 set heading off 时,在每页的上面不显示列标题,而是以空白行代替。

(5) 设置一行可以容纳的字符数。

SQL>SET LIN[ESIZE] {80|n}

如果一行的输出内容大于设置的一行可容纳的字符数,则折行显示。

(6) 设置页与页之间的分隔。

SQL>SET NEWP[AGE] {1|n|NONE}

① 当 set newpage 0 时,每页的开头有一个小的黑方框。
② 当 set newpage n 时,页和页之间隔着 n 个空行。
③ 当 set newpage none 时,页和页之间没有任何间隔。

(7) 显示时用 text 值代替 NULL 值。

SQL>SET NULL text

(8) 设置一页有多少行数。

SQL>SET PAGES[IZE] {24|n}

如果设为 0,则所有的输出内容为一页并且不显示列标题。

(9) 是否显示用 DBMS_OUTPUT.PUT_LINE 包进行输出的信息。

SQL>SET SERVEROUT[PUT] {ON|OFF}

在编写存储过程时,有时会用 dbms_output.put_line 将必要的信息输出,以便对存储过程进行调试。只有将 serveroutput 变量设为 on 后,信息才能显示在屏幕上。

(10) 当 SQL 语句的长度大于 LINESIZE 时,是否在显示时截取 SQL 语句。

SQL>SET WRA[P] {ON|OFF}

当输出行的长度大于设置行的长度时(用 set linesize n 命令设置),使用 set wrap on 命令,输出行的多余的字符会另起一行显示,否则会将输出行的多余字符切除,不予显示。

(11) 是否在屏幕上显示输出的内容,主要与 SPOOL 结合使用。

SQL>SET TERM[OUT] {ON|OFF}

在用 spool 命令将一个大表中的内容输出到一个文件中时,将内容输出在屏幕上会耗费大量的时间,设置 set termspool off 后,输出的内容只会保存在输出文件中,不会显示在屏幕上,极大地提高了 spool 的速度。

(12) 将 Spool 输出中每行后面多余的空格去掉。

```
SQL>SET TRIMS[OUT] {ON|OFF}
```

（13）显示每个 SQL 语句花费的执行时间。

```
SQL>SET TIMING {ON|OFF}
```

（14）遇到空行时不认为语句已经结束，从后续行接着读入。

```
SQL>SET SQLBLANKLINES ON
```

SQL＊Plus 中不允许 SQL 语句中间有空行，这在从其他地方复制脚本到 SQL＊Plus 中执行时很麻烦。比如下面的脚本：

```
select deptno, empno, ename
from emp

where empno='7788';
```

如果复制到 SQL＊Plus 中执行，就会出现错误。这个命令可以解决该问题。

（15）设置 DBMS_OUTPUT 的输出。

```
SQL>SET SERVEROUTPUT ON BUFFER 20000
```

用 dbms_output.put_line('strin_content')；可以在存储过程中输出信息，对存储过程进行调试。如果想让 dbms_output.put_line(' abc')；的输出显示为 SQL> abc，而不是 SQL>abc，则在 SET SERVEROUTPUT ON 后加 format wrapped 参数。

（16）输出的数据为 html 格式。

```
SQL>set markup html
```

在 8.1.7 版本以后，SQL＊Plus 中有一个 set markup html 的命令，可以将 SQL＊Plus 的输出以 html 格式展现。

13. 修改 SQL Buffer 中的当前行中第一个出现的字符串

```
C[HANGE] /old_value/new_value
SQL>l
  1* select * from dept
SQL>  c/dept/emp
  1* select * from emp
```

14. 编辑 SQL Buffer 中的 SQL 语句

```
SQL>EDI[T]
```

15. 显示 SQL Buffer 中的 SQL 语句，list n 显示 SQL Buffer 中的第 n 行，并使第 n 行成为当前行

```
SQL>L[IST] [n]
```

16. 在 SQL Buffer 的当前行下面加一行或多行

```
SQL>I[NPUT]
```

17. 将指定的文本加到 SQL Buffer 的当前行后面

```
SQL>A[PPEND]
SQL>select deptno,
  2 dname
  3 from dept;
DEPTNO DNAME
------- --------------
    10  ACCOUNTING
    20  RESEARCH
    30  SALES
    40  OPERATIONS

SQL>L 2
  2* dname
SQL>a ,loc
  2* dname,loc
SQL>L
  1 select deptno,
  2 dname,loc
  3* from dept
SQL>/

DEPTNO DNAME          LOC
------- -------------- --------------
    10  ACCOUNTING     NEW YORK
    20  RESEARCH       DALLAS
    30  SALES          CHICAGO
    40  OPERATIONS     BOSTON
```

18. 将 SQL Buffer 中的 SQL 语句保存到一个文件中

```
SQL>SAVE file_name
```

19. 将一个文件中的 SQL 语句导入到 SQL Buffer 中

```
SQL>GET file_name
```

20. 再次执行刚才已经执行的 SQL 语句

```
SQL>RUN
```
or

SQL>/

21. 执行一个存储过程

SQL>EXECUTE procedure_name

22. 在 SQL * Plus 中连接到指定的数据库

SQL>CONNECT user_name/passwd@db_alias

23. 不退出 SQL * Plus，在 SQL * Plus 中执行一个操作系统命令

SQL>HOST 操作系统命令

24. 在 SQL * Plus 中切换到操作系统命令提示符下，运行操作系统命令后，可以再次切换回 SQL * Plus，该命令在 Windows 下不被支持

SQL>!

SQL>!
$hostname
$exit
SQL>

25. 显示 SQL * Plus 命令的帮助

SQL>HELP

如何安装帮助文件：

SQL>@ ? \sqlplus\admin\help\hlpbld.sql ?\sqlplus\admin\help\helpus.sql
SQL>help index

26. 显示 SQL * Plus 系统变量的值或 SQL * Plus 环境变量的值

```
SHO[W] option
where option represents one of the following terms or clauses:
system_variable
ALL
BTI[TLE]
ERR[ORS] [{FUNCTION|PROCEDURE|PACKAGE|PACKAGE BODY|
TRIGGER|VIEW|TYPE|TYPE BODY} [schema.]name]
LNO
PARAMETERS [parameter_name]
PNO
REL[EASE]
```

```
REPF[OOTER]
REPH[EADER]
SGA
SPOO[L]
SQLCODE
TTI[TLE]
USER
```

(1) 显示当前环境变量的值。

```
SQL>Show all
```

(2) 显示当前在创建函数、存储过程、触发器、包等对象时出现的错误信息

```
SQL>Show error
```

当创建一个函数、存储过程等出错时,可以用该命令查看在哪个地方出错及相应的出错信息,进行修改后再次进行编译。

(3) 显示初始化参数的值。

```
SQL>show PARAMETERS [parameter_name]
```

(4) 显示数据库的版本。

```
SQL>show REL[EASE]
```

(5) 显示 SGA 的大小。

```
SQL>show SGA
```

(6) 显示当前的用户名。

```
SQL>show user
```

27. 查询一个用户下的对象

```
SQL>select * from tab;
SQL>select * from user_objects;
```

28. 查询一个用户下的所有表

```
SQL>select * from user_tables;
```

29. 查询一个用户下的所有索引

```
SQL>select * from user_indexes;
```

30. 在输入 SQL 语句的过程中临时先运行一个 SQL * Plus 命令

```
SQL>#
```

有没有过这样的经历？在 SQL*Plus 中输入很长的命令后，突然发现想不起某个列的名字了，如果取消当前的命令，待查询后再重输，那太麻烦了。当然，可以另开一个 SQL*Plus 窗口进行查询，但这里提供的方法更简单。比如说，想查工资大于 4000 的员工的信息，输入了下面的语句：

```
SQL>select deptno, empno, ename
  2 from emp
  3 where
```

这时如果想不起来工资的列名是什么了，只要在下一行以 # 开头就可以执行一条 SQL*Plus 命令，执行完后刚才的语句可以继续输入。

```
SQL>select deptno, empno, ename
  2 from emp
  3 where
  6 #desc emp
 Name Null? Type
 -----------------------------------------------------------
 EMPNO NOT NULL NUMBER(4)
 ENAME VARCHAR2(10)
 JOB VARCHAR2(9)
 MGR NUMBER(4)
 HIREDATE DATE
 SAL NUMBER(7,2)
 COMM NUMBER(7,2)
 DEPTNO NUMBER(2)

  6 sal >4000;

 DEPTNO EMPNO    ENAME
 ------------------------
 10 7839 KING
```

31. SQL*Plus 中的快速复制和粘贴技巧

（1）鼠标移至想要复制内容的开始。
（2）用右手食指按下鼠标左键。
（3）向想要复制内容的另一角拖动鼠标，与 Word 中选取内容的方法一样。
（4）内容选取完毕后（所选内容全部反显），按住鼠标左键不动，用右手中指单击鼠标右键。
（5）这时所选内容会自动复制到 SQL*Plus 环境的最后一行。

参 考 文 献

［1］ 何明. Oracle DBA 培训教程：从实践中学习 Oracle 数据库管理与维护［M］. 2 版. 北京：清华大学出版社，2009.
［2］ 腾永昌. Oracle 10g 数据库系统管理［M］. 北京：机械工业出版社，2005.
［3］ 陈俊杰，强彦. 大型数据库 Oracle 实验指导教程［M］. 2 版. 北京：科学出版社，2012.